伊能忠敬

TADATAKA INŌ AND HIS AUTOGRAPH

Tadataka Ino

the Japanese land-surveyor

by

Ryokichi Otani

Simon Publications
2001

First published in 1932: Iwanami, Tokyo

Translated from the Japanese by Kazue Sugimura.

Library of Congress Control Number: 33008136

ISBN: 1-931541-22-1

Printed by Lightning Source Inc. La Vergne, TN

Published by Simon Publications, P. O. Box 321, Safety Harbor, FL 34695

PREFACE

More than a century has elapsed since the preliminary coast survey of Japan was effectively carried out by an expert native surveyor. This is a marvel to those who consider the dawn of occidental civilisation in the country to have dated from the ratification of the treaty with different nations of the world. The opinion is, however, biased by the mistaken notion that science was not rooted in Japan until its introduction directly from the West. There is no doubt that the people felt the necessity of cultivating science after that time, but amateurs who aspired for the knowledge of various natural phenomena were not wanting even prior to that period. The superiority of western medical science over the Chinese was long recognized, and astronomy as well as geodesy was a subject in which many men were deeply interested.

The attention of many learned scholars was mostly centred in the elucidation of Chinese classics, which were but feebly linked with science; the study of nature, on the contrary, was rather held in contempt by such men as it is apparently contradictory to the doctrine of Confucius and his numerous followers. The schools founded by the feudal lords after the establishment of the Shogunate government were counted by hundreds, but the teaching of the Chinese classics was their principal aim. In spite of this imitation of the old Chinese method of instruction, there were a number of amateurs engaged in the development of mathematics and astronomy, though they received almost no encouragement from the side of the government. What the Shogunate desired was the construction of a right calendar, as theirs was sometimes found to be in error. The only astronomical observatory in the country was the one at Yedo; the observers absorbed only the rudiments of western astronomy, as the books they consulted were mostly written in Chinese. There were a few books in Dutch, which were scarcely understood because of the imperfect translation. The Chinese texts were compiled mostly by the hands of Jesuit missionaries, who were engaged in scientific pursuits while performing the duty of promulgating Christianity among the Chinese. Their writings were of infinite value to the Japanese amateurs, as they were easily intelligible, and gave the astronomical constants, the method of reducing the observations together with useful tables, including those of logarithms. The astronomers found immense treasure in these books, and the ephemeris was mostly compiled according to this storehouse of astronomical information. A Dutch translation of Lalande's Astronomie, which was imported towards the end of the 18th century, shed fresh light upon the cultivation of astronomy among a few Japanese scientists, who were eager to follow the progress of western civilisation.

Shortly before this period, there was an astronomer Gōryū Asada, whose originality in making stellar observations and whose methods of reduction far surpassed the other Japanese workers in the same field. He experienced various vicissitudes of fortune, and practised medicine for a livelihood in Osaka. His fame as an astronomer attracted the attention of eminent scientists in that big trading city. Prominent among them stood Gorobei Hazama, a pawnbroker; he was a well-known mathematician, and highly imbued with the idea of constructing precise astronomical instruments. A noted scholar was Sakuzaemon Takahashi, a petty municipal officer; he was endowed with high astronomical talent, had absorbed the fruits of Asada's long experience and was ready to remedy the defects of astronomical knowledge extant at that time, by his undaunted energy and zeal in pushing to the front the object at issue. Being promoted to an important position in the Calendar Office and the Astronomical Observatory at Yedo, he aspired to undertake *Gradmessung* in Japan.

The sphericity of the earth was known only to a small number of specialists at that time; any theory contradictory to the popular mind was considered as heretical, and severe punishment was inflicted on the promulgator. To frame any scheme, which had for its direct object the measurement of the length of the arc, was next to impossible. Doubtless the government officials would simply reject the proposal as useless. However, Takahashi was sagacious enough to take advantage of the confusion on the side of the government caused by the inroads of the Russians in one of the Kurile Islands nearest to Yezo. The motive of such raids was obscure—whether the Russians were intending to extend their territory or whether the marauders from Siberia came merely to ravage the unprotected islands; the need of mapping the coasts of Yezo was acutely felt by the officials, as the boundary between Japan and Russia was rather vague. This afforded a fine opportunity for surveying the northern frontier in a systematic manner, and if the work be accomplished, a part of Takahashi's ambition could be fulfilled. For conducting the field-work he found an ardent and experienced surveyor in one of his pupils, Tadataka Inō.

Inō was by trade a brewer; he accumulated a fortune by his assiduous toil and clever commercial tactics, after tasting all the bitterness of poverty. Fifty years old he was bold enough to call at the door of Takahashi, who was nineteen years younger, become his pupil and learn the art of surveying. His opulent means allowed him to construct various surveying instruments under the auspices of Hazama and by the hand of his skilful mechanician Ōno. In five years the brewer was metamorphosed into an experienced surveyor, who offered his services to make the map of Yezo at his own expense. This start was rewarded with excellent results, which called forth high appreciation of the officials, and were much to the satisfaction of Takahashi, who recommended Inō to the government for undertaking the complete survey of Japan. After sixteen years of indefatiguable toil and labour he brought the stupendous work to a close.

The story is told that the British Navy began to survey the coast of Japan in 1861, a few years after the ratification of the treaty between the two nations, but finding exact coincidence with the map of Inō, the survey was abandoned. Until the commencement of the Ordnance Survey with modern appliances, Inō's map gave the most reliable account of the geographical position of the different parts of Japan. It is wonderful how a small group of Japanese scientists accomplished such feats of geodetic measurements in the days when Japan was considered a hermit nation, barred from the introduction of accurate scientific knowledge cultivated in Europe.

Desirous of knowing the methods of observation and of reduction, the instruments, and the accuracy of measurements, I requested the Imperial Academy to collect the materials of the survey, since they would be effaced in course of time, and to compile a work on the deeds of the illustrious surveyor. The request was granted and the heavy task imposed on Mr. R. Ōtani, now professor of geodesy at the University of Kyōto. After seven years of untiring work, he brought to light many precious relics, and with his thorough knowledge of geodesy, analysed the results of the survey, examined the instruments, detecting slight misconception in their uses and in the method of projection.

The important discovery by Prof. Ōtani was the diary written by Inō himself, describing all the details of the survey from beginning to end. Many of the instruments are still preserved by his descendants; numerous original maps were still intact while the work of compilation was going on, but were unfortunately lost during the Tōkyō conflagration following the disastrous earthquake of September 1, 1923.

An account of Inō's life and work was published by the Imperial Academy in a quarto volume (xxxv+766) in May, 1917, written in Japanese. At the request of the Yamato Society, Prof. Ōtani has compiled an epitome of the book, which has been rendered into English by Mr. K. Sugimura, in order to present it to foreign readers. It is much to be regretted that in spite of the abundance of literature written in foreign languages on things Japanese, there is almost none illustrating the development of science before the opening of the country to foreign nations.

The present volume is published by the Yamato Society, whose object is to make clear the meaning and extent of Japanese culture in order to reveal the fundamental character of the nation to the world. The description of the coast survey at the beginning of the last century was deemed worthy of such publication, as it will no doubt prove the high aptitude of the Japanese for science even in the days when there was no encouragement for its study nor any recompense for its pursuit. As I took the initiative in collecting and analysing the works of Inō, I have added a few pages as preface, at the desire of the Society.

H. Nagaoka

The Institute of Physical and
Chemical Research, Tokyo
January, 1932

CONTENTS

INTRODUCTION

It is only since 1868, when the Meiji Restoration took place, that various sciences which had developed in Europe and America, together with the arts and customs of these countries, came to be introduced freely into Japan, and that the sciences were properly taken up and studied by the Japanese. Sixty years have hardly elapsed since then. From the standpoint of the Westerners, therefore, who used to know and hear of Japanese as folks wearing top-knots on their heads and carrying two swords at their sides, and with customs entirely different from their own, it must certainly be a marvel that the people, during the course of a half century, should have absorbed western civilization so thoroughly and made such marked progress in government and social improvement, in the sphere of science, and in that of industry. This, however, is no accidental miracle. As we look back on the past, four hundred years had already passed since Japan had come into contact with Europe. The interrelation between them was not new. Only as a consequence of the law prohibiting the people's welcoming western civilization freely during the Tokugawa Shogunate period, it appeared that Japanese culture was not appreciably influenced by western civilization. Some fragmentary knowledge of European sciences, however, was secretly introduced into Japan. Especially under the reign of such a patron of learning as the Shogun Yoshimune Tokugawa, a study of European astronomy was greatly facilitated and encouraged, and, besides, not a few books and instruments necessary for such study were imported. The seeds of knowledge newly imported from the West were gradually cultivated by our eager students, and the way had been thoroughly cleared even towards the end of the Shogunate régime for a ready reception and digestion of those western sciences which overwhelmingly flowed into Japan after the Meiji Restoration.

The achievements of land-surveying by Tadataka Inō, with which the present work deals, belong to the beginning of the 19th century—that is, sixty or seventy years before western sciences were formally introduced; and yet, in his method, there is a marked interfusion of some fragments of western sciences with those sciences that had already developed in Japan. This book should be not only a personal biography of Inō, but also a document concretely illustrating to what extent western sciences had been introduced at the time when Inō was active, and to what extent the native scholars were capable of utilizing them. In order to make clear the scientific conditions in the Japan of his time, we shall now proceed, before going into Inō's life, to a brief account of the development of astronomy, land-surveying, mathematics, etc. in Japan, entirely ignoring such sciences as medicine, botany, and those others which have nothing to do with land-surveying.

CHAPTER I

THE HISTORY OF THE DEVELOPMENT OF THE STUDY OF ASTRONOMY, MATHEMATICS, AND LAND-SURVEYING IN JAPAN

1 The Introduction of Chinese Civilization in Ancient Times

Although it is impossible to gain a knowledge of the real condition of learning in our ancient times, the incoming of Buddhism in the sixth century no doubt marked a new epoch in the development of our culture. For at that time the Crown Prince Shōtoku devoutly embraced the new religion and set about building Buddhist temples, and consequently a large number of technical experts came over from the continent, and with them various branches of culture which had developed previously in China.

Some Chinese calendars also were brought over to our country about this time. It was doubtless in accordance with Chinese custom that the Emperor Tenchi constructed a water-clock in 660 A.D. when he was Crown Prince, and that in 671 he appointed an official in charge of the water-clock for permanent timekeeping. In 692, a Chinese calendar was formally adopted in our country. This was the "Yüan-chia Calendar" designed by Ho-ch'êng-ti'en of the Tsung dynasty. This again was replaced in 697 by the "I-feng" or "Lin-tê Calendar," designed by Li-ch'ung-feng of the T'ang dynasty. The Taihō Statute, promulgated in the 2nd year of Taihō, or 702, provides for one master of astronomy, one master of calendars, one master of mathematics, two masters of water-clocks, ten assistants respectively in the departments of astronomy and calendar, thirty assistants in the department of mathematics, and twenty attendants in the department of time-keeping. It may be inferred from this that, perhaps merely for the sake of form, the officials were engaged in astronomical observations and calendar-making after the Chinese fashion. Then in 764, this calendar was replaced by the "Ta-yen Calendar" founded by Priest I-hsing of the T'ang dynasty. This, in turn, was replaced in 858, by the "Wu-chi Calendar" of Kuo-hsien-chih of the T'ang dynasty, and then this again gave place in 862, to the "Hsüan-ming Calendar" founded by Hsü-ang of the T'ang dynasty. The reason why calendars were so frequently adopted and displaced was that they were all inaccurate and their errors became so great in the course of time that the old calendars were discarded as soon as new ones were obtainable. Ever since the first introduction of Chinese civilization into Japan, envoys had very frequently been sent to China, and, besides, such a large number of Japanese students had studied in China that Japanese culture had attained marked progress, and literature and fine arts peculiar to the race had developed. In the study of astronomy, however, there was no such progress attained; the Japanese indiscriminately imitated the Chinese, adopting the Chinese calendar system with no rectification whatever. To make matters worse, the Japanese students of astronomy, after the tenth century, neglecting their proper field of study, drifted towards astrology and gradually came to lack the ability of calendar-making altogether. Besides, the eleventh century witnessed the commencement of incessant civil wars, and the connection with the continent was not so intimate as before. For the following several centuries, learning was completely neglected and the officials in charge of calendars were inefficient, helpless before the large errors existing in their calendar, which might be easily detected even by a rough astronomical observation. They paid no attention to new calendar systems even though some were brought over. Deplorably dark ages ensued, when calendar dates were carelessly fixed according to the antiquated Hsüan-ming Calendar system.

With these Chinese calendars also came Chinese mathematics. Although the Taihō Statute mentions the "Chou-pi-Arithmetic," the "Chiu-chang-Arithmetic," and some other works on mathematics, we cannot decide to what extent mathematics was developed in Japan at that time. We may suppose that it was studied enough to make possible simple calculations in connection with calendar-making and daily business. It seems that mathematics was not much studied as a science.

Concerning the method of land-surveying in our ancient times also, we can only venture very vague guesses. It is evident that a magnetic compass was imported from the continent as early as the 6th century. Moreover, the presence of such a title as the "Hai-tao-suan-shu" (Sea-Island Arithmetic) among the text-books mentioned in the Taihō Statute indicates that some knowledge of land-surveying had been brought over from China together with astronomy and mathematics from the earliest days. Again, judging from the existent traces of a systematic allotment on the plains of the Kinai region according to a law promulgated sometime in the 7th century, and, also, from the statute governing the extensive street plans of the two cities—Nara, built at the beginning of the 8th century, and Kyōto at the end of the same century—it may be inferred that a method of surveying was put into practice at that time. The area on which the method of land-surveying was really effectively applied was no more than 10 kilometers square, and there is no evidence that an attempt was made over a more extensive area with a view to map-making. Gyōgi was a great Buddhist priest in the 7th and 8th centuries, who went about the country, building a large number of roads, bridges, canals, and embankments. He was a veritable civil engineer. The oldest map made in Japan is attributed to this priest. This map was extremely crude and primitive, roughly representing the shape of our country. The important art of land-surveying made no marked progress before the 10th century, when it was destined to decline with other sciences.

2 The Revival of Mathematics and Adoption of a New Calendar in the Jōkyō Era

When the sciences began to decline, mathematics was so thoroughly neglected that there existed but few people even with the knowledge of division. In the early years of the 17th century, however, mathematics was suddenly revived before any other science. At the close of the 16th or the beginning of the 17th century, a man named Shigeyoshi Kambei Mōri came across a mathematical work entitled "Suan-fa-t'un-tsung" written by Ch'êng-ta-wei of the Ming dynasty. This book expounded the method of calculation by means of a *soroban* (abacus). Mōri wrote a book on the basis of this work and spread the art of abacus calculation. Mitsuyoshi Yoshida, one of his pupils, wrote a book entitled "Jinkōki" for the purpose of popularizing the art. Although likewise based on the Suan-fa-t'un-tsung, the Jinkōki was a genuine Japanese version of it, expounding the method of using an abacus in the four operations of arithmetic, and in extracting square and cube roots, and so on, and proved of great service in daily life. As the work was published in 1627—that is, more than twenty years after the establishment of the Tokugawa Shogunate—a time when the nation was singing the praises of peace, it was enthusiastically received and immediately became quite popular. Another book which, with the Suan-fa-t'un-tsung, proved instrumental in the revival of our mathematics is the "Suan-hsüeh-chi-meng," written by Chu-shih-chieh of the Yüan dynasty. It was this book that introduced to our country the "T'ien-yüan method" or the Japanese *Tengen* method. In this the operations were carried on by means of the calculating pieces called *sangi*. The date when this original work was imported is not clear, but an edition of the work with Japanese pronunciation was published about the middle of the 17th century, and foreshadowed the revival of mathematics. With the appearance of Kōwa Shinsuke Seki (1642–1708) on the stage, rapid strides were made in our mathematics, and the foundation of the so-called *wasan* or Japanese mathematics was laid.

Seki, a mathematical genius, though sometimes called a pupil of Yoshitane Takahara who studied under Mōri, was a self-taught scholar who developed his unique talents. He first improved the *Tengen* method and designed the *Tengen-Endan* method, which he later on developed into the *Tenzan* method. In the *Tenzan* method the relations which exist between various quantities under considerations were to be expressed in writing and the analysis leading to the final results was also to be carried out by writing, somewhat similar to the algebra of western countries. Then he invented the *Shōsa* method, the *Tetsujutsu,* and others, and finally the *Yenri* method, which is, in a sense, similar to the differential and integral calculus, and may be used in calculating the length of curves, areas, and volumes.

It is true that our mathematics attained an unusual advance with the appearance of Seki. However, the methods he invented were by no means perfected by his efforts alone. It was a custom in the Tokugawa period that the more profound secrets which the masters had attained should not be made public, but be transmitted to only a few of their pupils. Therefore, the very heart of Seki's mathematics was likewise pursued and studied by such mathematicians of his school as Murahide Araki (1640–1718), Masahiro Takebe (1664–1739), Yoshisuke Matsunaga (?–1746?),

Genkei Nakano (1662–1733), Shujū Yamaji (1704–1772), Yoshita Kurushima (?–1757), Chokuyen Ajima (1739–1798), Sadasuke Fujita (1734–1807), Makoto Kusaka (1764–1839), Nei Wada (1787–1840), etc. The Tenzan method, for instance, was for the first time made public in 1766 by Raidō Arima, Daimio (Feudal Lord) of Kurume in Chikugo, as he discussed it in his "Shūki-sampō," thus entirely breaking all traditions. Such a secret as the Yenri method had never been brought within reach of outsiders before the Meiji Restoration. As it was pursued by students, few in number, but with keenest intellect, steady development resulted. Especially the efforts of Chokuyen Ajima brought about marked progress, and again Nei Wada improved the method with remarkable cleverness. It must be noted that, though the so-called *wasan* is generally considered synonymous with the mathematics of the Seki school, there have been not a few scholars outside that school who have contributed to the development of Japanese mathematics. Yasuaki Aida (1744–1817), for instance, though educated for some time in the Seki school, opposed his former master, and established a new school where he gave expression to his rare mathematical genius. Among the many others outside the Seki school, the Takuma school, founded by Yoshikiyo Takuma, was popular in and around Ōsaka and produced a group of very able mathematicians. Here it was that the mathematical training of Yoshitoki Takahashi and Shigetomi Hazama was given—the two scholars who inspired Tadataka Inō to accomplish his unprecedented enterprise and also rendered a valuable contribution to the advancement of the study of astronomy, with which we shall deal later.

As mathematics was thus revived, the study of astronomy gradually awoke into activity from its prolonged slumber. It goes without saying that knowledge of mathematics is indispensable in calendaric calculations. Especially during a period when practical observations were scarcely made, the study of calendars was conceived as a branch of mathematics. Therefore, it was only natural that one should have shared the fate of the other. The "Hsüan-ming Calendar" had been continually used since the middle of the 9th century, but during the 17th century it was found that the dates given in this calendar were in error by almost two days compared with the actual season; and the forecasting of eclipses by the calendar turned out so frequently to be false that the people at large discussed the subject freely and keenly felt the importance of studying calendars for themselves. Shōi Ikeda, a mathematician of the Seki school, having access to the "Shou-shi Calendar" designed by Kuo-shou-ching of the Yüan dynasty, thoroughly mastered it. However, it was withheld from the public, and remained a secret strictly preserved among his pupils. Among Ikeda's followers, there was one named Shunkai Sukezaemon Shibukawa (1639–1715). He was formerly called Santetsu Yasui, a professional *go* player employed by the Shogunate. This man, being naturally fond of astronomy, was engaged for many years in observing the heavenly bodies, and designed a new calendar which coincided with actual heavenly phenomena. He then appealed to the Throne more than once to have his calendar accepted. The officials in charge of calendars at Kyōto had also realized the necessity of a new calendar, and in 1684 they attempted to compile a new one based on the "Ta-t'ung Calendar" designed by Yüan-t'ung of the Ming dynasty. The Ta-t'ung Calendar is almost identical with the Shou-shi Calendar except in the matter of the periodical variation of the length of year. Thereupon, Shibukawa appealed to the Throne again, strongly protesting against the absurdity of adopting, with no modification whatever, a calendar designed, not for Japan, but for a foreign country. An observatory was established at Kyōto to be used for observing the motions of the heavenly bodies.

Testing various calendars by the observations it was finally decided to adopt Shibukawa's new calendar, or the "Jōkyō Calendar" as it was graciously named by the Emperor, Jōkyō being the name of that era. This calendar, though likewise based on the Shou-shi Calendar, had been designed according to Shibukawa's own observations and the original calculations which he had worked out, taking Kyōto as the origin. It was entirely different from those which had been brought over from China and employed with no modification. The Jōkyō Calendar may be called, in the strict sense of the term, the first Japanese calendar proper.

As the Jōkyō Calendar, based on the Shou-shi Calendar, was adopted as the state calendar, not a few mathematicians commenced to study the original and wrote a good many books upon it. For instance, Masahiro Takebe and Genkei Nakane were not only able mathematicians, but also calendaric scholars thoroughly acquainted with the Shou-shi Calendar. Since the adoption of the Jōkyō Calendar, the authority of compiling a calendar was practically transferred from the calendar officials at Kyōto to the Shibukawas, astronomers of the Shogunate at Yedo. This hereditary profession had the usual defect, since not all the descendants of Shibukawa were fit to be calendar scholars. When the time came, in the middle of the 18th century, that the Jōkyō Calendar was felt to need some amendment, the astronomical officials of the Shogunate had lagged so far behind the progress of sciences that they were unfit for the important duty of compiling a new calendar. Thereupon, an attempt was made to employ men fairly well acquainted with western astronomy which had come over into Japan by this time, and to introduce some western knowledge into the Japanese calendar. But this effort failed. Then, enlisting the assistance of Shujū Yamaji and other mathematicians, they at last managed to compile a new calendar in 1754. But their unfortunate lack of care in observation and calendar-construction betrayed itself within a few years when gross errors appeared in the computation of solar eclipses. So some minor corrections were applied to the calendar in 1769. By that time there were, among the common people, a number of students who were capable of mastering and digesting the Shou-shi Calendar and other calendars imported from China. Some people even set about studying western calendars. The study of calendars, neglected by the officials in charge, promised to prosper among the people. When the official calendar miscalculated a solar eclipse in 1763, quite a number of private calendar scholars had pointed this out beforehand.

3 The Introduction of European Sciences

It was in the year 1541 that Europeans came to Japan for the first time. In that year a typhoon blew a Portuguese trader to the shores of Bungo. Another trader from Portugal drifted to Tanegashima in 1543, and then it was that rifles were introduced to this country for the first time. This was towards the end of the Ashikaga period—a time when, on the one hand, the native warriors all coveted such new powerful arms; and, on their part, the Portuguese were anxious to extend their religion and trade. They had already occupied Java in the South Seas, where they had preached Catholicism, and they had opened trade with China. Therefore, communication with Portugal was opened quite speedily. Englishmen followed suit in 1580, Spaniards in 1581, and Dutchmen in 1597. Catholicism was introduced in 1549, and within a few years it spread so rapidly in Kyūshū that not a few feudal lords were converted. Trade and religion steadily prospered side by side. On Nobunaga Oda's assumption of power, he granted the construction of a Catholic church named "Nambanji" in Kyōto for the purpose of preaching the new religion. It now spread with tremendous force from the Kinai and Chūgoku districts even to those of Kwantō and Tōhoku. However, as soon as Hideyoshi Toyotomi began to rule over the country, succeeding the Oda family, he recognized certain dangers in the Christian propaganda and issued a proclamation against it. Trade with foreign countries was not forbidden, though, and foreigners were permitted to come and go so freely that the national interdict against Christianity could not be enforced sufficiently to prevent its spread. When Iyeyasu Tokugawa established the Shogunate, he followed Toyotomi's original policy to prohibit Christianity and to grant trade privileges alone. His injunction, however, like that of his predecessor, was not well observed. Besides, the attitude assumed by Portugal and Spain towards Japan proved to be one of aiming at the invasion of the country through the medium of evangelization. Therefore, the Shogunate in 1612 proclaimed a strict interdict against Christianity all over the country with a view to preventing the possibilities of its propagation. The Christians gathered in revolt at Shimabara in Kyūshū in 1637 under the banner of the Cross. The law was even more drastically enforced after the suppression of the riot in the following year; an isolation law barring all communications with European countries was issued, the only exception being Dutch merchantmen which were permitted to come for trading.

From the first introduction of Christianity into Japan, most of the foreign missionaries were skilful physicians, versed in astronomy, geography, and the art of fortification. They won the hearts of the people by medical treatments and endeavoured to please feudal lords by imparting bits of the art of fortification. The introduction of European sciences into Japan, therefore, may be said to coincide with the incoming of Christianity. The aim of the missionaries was of course the propagation of their creed, but not that of sciences. Moreover, because incessant civil wars afforded the people no leisure to study pure science, the missionaries introduced European sciences merely as a means of the work of evangelization; so seldom, if ever, did they teach medicine, astronomy, or other pure sciences to the Japanese, as it seems. The influence of Christianity upon the minds of the Japanese during the 16th century must have been tremendous, while that of European sciences upon the evolution of the native sciences was extremely slight—hardly worth mentioning. When the 17th century brought peace and quiet, so that the nation might again turn to arts and sciences, they had no access to European sciences through missionaries because of the prohibition of Christianity, nor through reading because of the law decreed in 1630 originally prohibiting books on Christianity, but soon enforced so drastically as to ban the importation of books not only on religion but on any subject in which such terms as Jesus, God, western countries, or the names of missionaries might be found. Books of that day treating of western sciences were, as a rule, those translated into Chinese by missionaries stationed in China and were necessarily full of such tabooed words. Therefore, books on astronomy, geography, mathematics, etc. were ruled out by this law. But the progressive spirit of the times could not be suppressed and kept away from the western sciences. Just as Christianity always thrived under such rigid prohibition, the western sciences were clandestinely spreading. In 1720, the Shogun Yoshimune Tokugawa, who had a keen interest in astronomy, modified the injunction so as to permit the importation of books on astronomy, geography, mathematics, etc., and also in 1745 granted permission for the interpreters at Nagasaki to study Dutch books. Henceforth, the Japanese had access to western sciences openly through Chinese translations as well as through the few original books brought over from Europe. But so rare and valued were these works that it was extremely difficult for common people to come within reach of them, and even eager scholars, unless in office, experienced extraordinary difficulty in obtaining a chance to read them. We shall now proceed to discuss what influence these western sciences had upon the Japanese sciences from the beginning of communication between the East and the West down to this period.

(*A*) *Mathematics*

Whether any western mathematics was introduced into Japan during the 16th century when Christianity was most flourishing cannot now be determined. However, no influence of western mathematics can be traced in the course of development which took place between Shigeyoshi Mōri's abacus calculation and Kōwa Seki's *Tengen* method or his *Yenri* method. Even after the ban on foreign books had been somewhat loosened and Chinese translations of western mathematics were imported, the direction of the development of Japanese mathematics can hardly be said to have been affected by foreign works. The reason probably is, that Japanese mathematicians had already advanced in their unique form and in some parts even beyond the Westerners by the time the repression was mitigated. For instance, the Japanese had already mastered the numerical solution of equations of higher degree when such was not yet known in western countries. The idea of determinants had also been conceived and utilized by Seki in his calculations, even before Leibnitz. It seems that the Japanese scholars were not much impressed by what they found in the foreign books, most of which were Chinese translations, and that they were convinced that their mathematics was superior to that of Europe. Only trigonometry, the trigonometrical table, and the logarithmic table which appeared in the foreign books were quite new to the Japanese. These were at once adopted for use in calendaric calculations. Above all, the origin and calculative method of the trigonometrical table were speedily expounded by means of Japanese methematics, and a great deal was written on the subject. The logarithmic table and its uses were mentioned in a Chinese book entitled "Shu-li-ching-yün" and in others, but nothing was said concerning the process by which the logarithms were arrived at. Not a few Japanese assiduously laboured in working out the process. At length Chokuyen Ajima succeeded in inventing a method while Yasuaki Aida also contrived a quite different one. This was done only through curiosity on the part of the Japanese mathematicians, and it does not indicate any change in their ideas. As for trigonometry, they simply adopted it for its practical use, but there is no evidence that they investigated it seriously. The European style of geometry seems to have been entirely ignored by the Japanese. In brief, European mathematics introduced before the end of the 18th century failed to obtain serious consideration from the Japanese mathematicians in spite of its partial contributions to calculations in calendar-compiling and land-surveying. Possibly it gave some new hints to the native scholars, but, on the whole, failed to exercise any appreciable influence upon Japanese mathematics. It may be said, therefore, that Japanese mathematics as such was making distinct progress.

(*B*) *Astronomy*

As to when European astronomy was introduced into Japan for the first time it is unknown. But it is evident that it was brought over before the middle of the 17th century, when the Japanese students commenced a study of the Chinese Shou-shi Calendar. There existed in 1659 a book entitled "Kenkon-bensetsu" (Discussions of the Universe) written by Genshō Mukai and others. The original of this work was brought over by some Christians who came adrift to the shores of this country in 1643, and was first translated into Japanese in Roman script by Chūan Sawano (originally, Christovão Ferriera), a Portuguese who was naturalized as a Japanese. Mukai translated this work into Japanese writing with some notes. This book attempts to explain earthquakes, tides, meteors, thunder, rainbows, and all other phenomena (except a few matters such as the periods of the revolutions of the sun, the moon, and the five planets), all by means of the correlations of the four elements—namely, water, fire, air, and earth. It is by no means a treatise of pure astronomy, nor has it much scientific significance. However, it is the oldest of the books dealing with western views, and serves to prove that such an idea had been introduced into Japan by this time. Tradition says that during the first half of the 17th century there lived in Nagasaki an interpreter named Kichizaemon Hayashi, who was well versed in western astronomy and was executed in 1646 on suspicion of being a Christian. Yoshinobu Kobayashi (1601–1683), his pupil, after being imprisoned for twenty-one years, was released in 1667, and this man is responsible for the serious study of western astronomy in Japan. How much knowledge these men possessed is not clear. Only Kobayashi's scholarship may be guessed at from the fact that he called attention beforehand to the miscalculation of a solar eclipse announced in the government calendar of 1683. Joken Nishikawa (1648–1724) and his son, Seikyū Nishikawa (1693–1756), both of Nagasaki, are said to have been masters of the Dutch language and well versed in western astronomy. Among Joken's works there exists a commentary on the "T'ien-ching-huo-wen," an ancient Chinese book introducing western theories of astronomy. Seikyū also rendered a Japanese version of this same Chinese work. The former, in 1719, was called to Yedo by Yoshimune to answer some questions in astronomy. The latter was appointed official astronomer to improve the Jōkyō Calendar, introducing some aspects of western astronomy into it, a task which, however, failed of completion owing to the opposition of the rival officials at Kyōto. The residents of Nagasaki had access to foreign books and foreigners and, therefore, western astronomy made some development first in this part of the country. And yet in the earlier part of the 18th century, there was no one who thoroughly understood western theories. Even after the interpreters at Nagasaki were permitted to learn the Dutch language freely, not one appeared in this region who mastered western astronomical theories, partly because original works on the essentials of western astronomy were not introduced before the end of the 18th century, and partly because none of the interpreters had any taste for such a branch of science. However, we should not fail to record here the fact that Tadao Shidzuki in 1798 wrote the "Rekishō-shinsho" (A New Book of Astronomical Phenomena) on the basis of John Keill's work, and as an appendix to his book, Shidzuki gave his own opinion which resembles the nebular hypothesis.

Masahiro Takebe and Genkei Nakane were scholars of calendars as well as of Japanese mathematics who were favoured by the Shogun Yoshimune. These scholars are commonly said to have influenced the Shogun Yoshimune to mitigate the ban on foreign books. They had read part of the "Li-suan-ch'üan-shu" on the western theories compiled and expounded by Mei-wên-ting and found it to be a useful work; and, it seems, they had persuaded the Shogun to collect all Chinese translations and treatises on western astronomy. Already in China the "Ch'ung-chêng-li-shu," —a compilation of western astronomical theories and methods used in calendaric computations, had been written in 1636 by missionaries such as Giacomo Rho and Adam Schall; and the "Ling-t'ai-i-hsiang-chih" which illustrates western astronomical instruments had been compiled in 1674 by such missionaries as Ferdinand Verbiest. There is no room for doubt that these books came into the library of the Shogunate at the latest along with the mitigating of the ban on foreign books in 1720. During the Kyōho era (1716–1735) some carpenters of Nagasaki, according to an order from the Shogun Yoshimune, constructed a kind of meridian circle called *Sokugohyōgi.* The principal part of this instrument was a quadrant, equipped with a telescope, of a 1·2 m. radius—a very fine piece of work set up in western mode. Yoshimune's death brought the pursuit of this study in official quarters to a stand-still, and as a revision of the calendar was planned in 1754, the calendar officials at Kyōto and the astronomers of the Shogunate, and a few private scholars of calendars invited for co-operation, were commanded to make use of the newly introduced theories; but save for the use of a few instruments in western style like a quadrant in making observations, and the trigonometrical table in computation, the essentials of calendaric theories remained as old-fashioned as ever. Then also the "Li-hsiang-k'ao-ch'êng" which treats of western astronomical theories, and the "I-hsiang-k'ao-ch'êng" which gives the positions of the fixed stars, etc., must have been imported into the country as soon as they were published, and must have entered the Shogunate library first of all. It is a strange phenomenon that while the astronomers of the Shogunate who had ample access to these books introducing western theories almost altogether failed to study them, private students of calendars who had the greatest difficulty in gaining access to them, gradually developed the study through their own untiring efforts. Gōryū Asada was one of such eager scholars who attained his purpose and marked a new epoch in the development of astronomy in Japan.

(*C*) *Land-Surveying*

As stated above, some fragmentary method of land-surveying came over from China in ancient times and was applied in various ways, but it was almost lost, as it were, during the later centuries. Therefore, when the method of land-surveying was introduced from Europe, it seemed as though this study were resuscitated by the western method. According to a record of our land-surveying, the western method of land-surveying was first introduced during the Kwan-ei era (1624–1643) as a man called Gon-emon Higuchi of Nagasaki was taught this method by a Dutchman by the name of Casper. It is said that as Higuchi for seven long years persistently implored the Dutchman to teach him this method, the latter was finally moved to consent by the former's earnestness. The times were such that, as peace reigned again, exact land-surveying was pressingly demanded for map-making, constructing water-works, and building canals. Consequently, as soon as Higuchi mastered the western method of land-surveying, his house was thronged with people asking for instruction. Sadanaga Shimatani, Unsetsu Hirai, Kyūya Yamasaki, and Gyōbuzaemon Kanazawa all completed the full course offered by Higuchi. As the method of land-surveying was popularized, the people, being struck with its wonderful ingenuity and convenience, began to regard it superstitiously to such a degree that the Shogunate deemed it wise to prohibit the teaching of this method to common people without permission. But the authorities could not long suppress the popularization of such a useful study. Their prohibition became less and less severe till the study was pretty freely popularized. During several decades, however, it persisted only in secret and its development was considerably checked. Gyōbuzaemon Kanazawa handed down the method of land-surveying he had studied under Higuchi to his son, Seizaemon Kanazawa (1624–1684) who taught it to his younger brother, Kan-emon Kanazawa (?–1691), and he in turn gave it to a man by the name of Sadanori Taemon Shimidzu (1646–1717). This man, fearing that the method might be entirely lost, deliberately arranged and enlarged what he had studied under his master, and prepared, in 1694, the prohibition then being lifted, "Kikujutsu-honden," "Betsuden," and "Inka-no-maki" (The Main Course of the Method of Surveying, Higher course, and Further Studies) for his pupils. At last he went up to Yedo in 1712 and founded a large school, stirring up a considerable revival of the study. His has generally been called the "Shimidzu School," and most students of land-surveying of later days followed this Sadanori Shimidzu to some extent. Another group following Kyūya Yamasaki established themselves as the "Yamasaki School," which also boasted of a great number of students after the lightening of the prohibition. Still more time was required before publishing their reasonable methods in books. For instance, the "Jishakusan-kon-genki" written by Yoichiemon Hosaka in 1687 may be considered as the oldest publication, though fragmentary, along the line of land-surveying, but its contents are so poor that it is hardly worth mentioning.

A work on land-surveying of some importance is the "Kikubuntō-shū" written in 1722 by Tokiharu Mano. In this book an illustration is given by a diagram of a brass surveying-instrument, equipped with a magnetic compass—a kind of alidade. This instrument is so constructed as to serve for plane-table survey by the method of traverse or the method of intersection, and also for measuring altitudes. But throughout the book no mention is made of this appliance as though this had been kept from publication. The surveying methods expounded in the work are only round-about, such as making use of the diagram of a right-angled triangle, and they by no means correspond with the instrument introduced. Levelling is suggested as being carried out by means of the level of the water run into a narrow groove cut in a piece of wood. Simple methods of measuring the height of mountains and the width of rivers are given. A compass with a circle graduated in the reverse order is illustrated by a diagram, but no mention of its use accompanies the diagram. In short, this work, though full of fragmentary items, fails to give a consistent method of land-surveying.

The next book on land-surveying is the "Ryōchi-shinan" (A Guide-Book of Land-Surveying) written by Masahiro Murai. The work is divided into two parts, the first of which was published in 1733. In the volume, surveying-instruments are illustrated, such as a plane-table to be employed for both horizontal and vertical positions, a straight rule to be used as a simple alidade, a plumb for horizontal and vertical tests, a measuring-rope, a measuring-pole, a sign-post, a telescope, drawing-compasses and protractor for map-making, etc. Processes of surveying by using these instruments are stated. The processes substantially correspond to the so-called plane-table survey of to-day which is chiefly executed by the method of traverse and the method of intersection. The plane-table was so set up at the first station that one side of the table coincided with the direction of the line to be measured, and at the later stations the orientation of the plane-table was adjusted solely by the direction of the line already measured, without relying upon a magnetic compass at all, and so the method adopted here is one which never relied upon a magnetic compass. The second volume compiled in 1754 explains different uses of the magnetic compass. The graduations of a compass circle may be either direct or reverse. The use of magnetic theodolites of different sizes is explained, and a general statement of land-surveying with the use of a magnet is given. It also describes a simple distance-finder called *hōjaku* (cheek measure), and a perambulator, and also touches the method of levelling. Compared with the first, the second volume lacks unity, but to those who wished to educate themselves in land-surveying, the second volume as well as the first must have been very helpful at this early period.

The "Choken-bengi" (Discussions on Surveying) written by Dōkan Shimada in 1734 discusses different people's comments on land-surveying and gives the author's opinion. In the book the author describes a method called *enzan-ka-mokuteki* which means distant mountains as temporary sighting-objects. This is a method by which accuracy or faultiness in a survey can be proved: first, the azimuths of high mountains or other objects sightable from different positions are measured from various points while carrying on survey by the traverse method; and then, whether the azimuth lines for the same object should all meet at one point or not is tested on the map constructed with surveyed data. In the work mention is made of the possibility of determining one's positions at sea by measuring the latitude and azimuth with a quadrant, a vertical circle, and a magnetic compass. But as to its process nothing is said.

Besides these published books, there are some which have been copied and circulated, among which the following may be considered the chief: the "Kiku-gempō-hiden" (Secret Principles of Land-Surveying) by Sadanori Shimidzu, the "Chiiki-zuhō-daizen" (A Complete Handbook of Land-Surveying) by Chishin Hosoi, written in 1717, and the "Bundo-yojutsu" (Method of Surveying) by Toshiyori Matsumiya, written in 1728. The method of land-surveying mentioned in the Chinese works such as the Ch'ung-ch'êng-li-shu and the Li-suan-ch'üan-shu must also have been read by some of the scholars as the ban on foreign books had been mitigated.

Taking into consideration the development of land-surveying in Japan in the latter half of the 18th century from the contents of the works just mentioned, the following may be concluded. In surveying level land there existed methods exactly similar to the traverse, the offset, and the method of intersection of the present day; and the method of levelling then employed does not differ in principle from the one practised to-day. Proper steps were taken to test the accuracy of surveying and accumulation of errors was guarded against. As to surveying-instruments there were magnetic theodolites, sticks capped with magnetic compasses, and water-levels (spirit-levels were not yet imported) as well as several kinds of measuring-ropes, plane-tables, and alidades.

The fact that the magnetic compass does not point to the exact south or north was already known by the Japanese through foreign books, and this is mentioned in the calendaric work compiled on the adoption of a new calendar in 1754. Among the works on surveying, some discussed a safer process than setting up a plane-table at each station by a magnetic compass: namely, that of fixing the plane-table by back-sighting after once strictly adjusting its azimuth at a station. Thus the methods and instruments necessary for surveying a piece of land for which the surface of the earth might be treated as a plane were practically complete. However, the construction of the instruments was crude and hardly worth the name. The magnetic needle was especially unfit in shape and in manner of setting, and scarcely suited for a direction-pointer. The theodolite with too large a graduated circle, too roughly graduated, and with a short crude magnetic needle, did not prove of much help in practice, either. As to land-surveying treating the earth as a globe, mention is made only of the existence of a method of determining latitude by a measurement of the altitudes of the sun or circumpolar stars with a kind of quadrant. No document describes the practice of such a method. It seems evident that among the scholars of land-surveying there was none capable of making astronomical observations with a view to utilizing it in determining the position of stations. The so-called triangulation of the present day was naturally an unknown thing at that time.

Now let us consider to what extent these methods of land-surveying were practically employed. Seizaemon Kanazawa by order of the Shogunate surveyed the streets of Yedo and made a map of the city in 1657, which seems to be the first instance in which surveying was applied to a comparatively large area. Tradition says that there was a conflagration in Yedo that year which raged for two days and nights and a large number of people lost their lives. This disaster made the Shogun consider the lack of a good map of the streets as the cause of so many casualties, and immediately ordered a map of Yedo to be made. Hence expert surveyors were summoned, and Kanazawa was appointed for the task. He is said to have worked exclusively on the survey, and succeeded in making a map in several weeks. This map has been unfortunately lost. Not much later than this, in 1670 and during several succeeding years, a survey-map of Yedo was published in several sectional sheets by a certain man under the pseudonym "Ochikochi-Michijirushi" and this has been preserved until to-day. This is said to have been based on Kanazawa's survey-map. And it is rather surprising that of all existent maps of counties or villages constructed at the end of the 17th century or in the 18th century, so very few should have been reasonably constructed when it was quite easy to make survey-maps. Exact maps of such small areas were much needed by the feudal lords throughout the country, and even a simple method of surveying would have answered the purpose. Even crude surveying-instruments would not have occasioned great errors. The Shogunate's suppression of popularization of land-surveying for some time may account for such influence upon the practice of land-surveying.

There were several kinds of maps of the whole country compiled by the authorities: the first being the Bunroku maps which were constructed by feudal lords by the order of Hideyoshi

Toyotomi at the end of the 16th century; the Shōhō maps of the provinces, counties, and castles, which the feudal lords commenced by the order of the Shogun Iyemitsu in 1644 and completed after more than ten years; and the Genroku maps, the revised maps of the provinces set about in 1697 by the order of the Shogun Tsunayoshi and completed in five years. These maps all consisted of several dozen or even several hundred sheets, and the names of mountains, rivers, post-towns, and villages were written in considerable detail; but, of course, they were not properly drawn after surveying. Naturally their topographical representation was quite inaccurate, and considerable discrepancies often occurred between the maps of adjoining areas. Therefore, in 1719 the Shogun Yoshimune commanded Masahiro Takebe and others to carry out an azimuth survey with a view to revising the old maps, and thus finally succeeded in obtaining a map of the whole country. It is unfortunate that this map, though completed in 1723, has been lost. But Takebe's record concerning his method of map-making has been preserved to this day. According to this record, three suitable points were first chosen in each province; then the azimuths of the high mountains in the surrounding provinces were sighted from these points; then the old maps were reproduced on a reduced scale of 1:216000, and the azimuth lines worked out by practical survey were drawn on the reduced maps from the appointed points represented thereon; and maps of different provinces were pieced together referring to these azimuth lines, and in case they did not coincide on the boundaries, the difference was eliminated by cutting away the surplus and by supplementing the shortage. However, it should not be supposed that Takebe was completely satisfied with such a makeshift, because he writes in his postscript that in order to obtain an accurate map, it would be necessary to determine the latitudes of places by closely measuring the altitude of the pole star, and the longitudes by observing the lunar eclipses; at the same time to survey the circumference of each province and also the routes crossing the province lengthwise and sidewise by the traverse; to determine the position of the famous mountains by the method of intersection. It is evident that Takebe thoroughly understood the proper method of making an accurate map of an extensive area. Takebe's talents and encouragement from the Shogun Yoshimune, patron of learning, would have made it no difficult task to construct an accurate map of the whole country at the beginning of the 18th century. It is a great pity that no such effort was made.

Besides these maps compiled by the government officials, there were, in the latter part of the 17th century, several published maps giving the whole country on a single sheet of paper, and some others were copied and circulated. Especially the map alleged to have been drawn by Terutsuna Matsudaira, Daimio of Koga, in 1661 or a little before, is crude enough as a map, but it is remarkable that it is drawn in the form of the so-called compass-map employed in Europe, and it marks latitudes on its border. The Japanese maps drawn by foreigners down to the 17th century are really no maps at all; but after a voyage was made by Martin Gerritzoon Vries and others about the middle of the 17th century, the shape of the Japanese Islands became somewhat definite to Europeans, and some improvement is apparent in the maps published later. The map inserted in his book by Engelbert Kämpfer who visited Japan in 1690 is based on both Japanese and foreign materials. These maps were all crude and only showed a rough outline of the country. Even during the 18th century, such maps as were accessible to common people failed to make great progress; and those made in foreign countries were not much improved in spite of the fact that more and more foreign ships explored the Orient. The reason was perhaps that on account of the severe isolation law they were unable to survey the coasts of Japan.

Genshu Nagakubo, a native of Mito, in 1778 published a complete map of Japan commonly called the "Sekisui map"—Sekisui being the pen-name of Nagakubo. This was the first time that a fairly trustworthy representation of the land was produced. It was merely a compiled one based on old maps, but well revised according to his own experience gained by his extensive travel throughout the country. Besides, he took into consideration the scanty results of astronomical observations which some calendar scholars had obtained at various stations. As it was by no means a survey-map the positions and shapes of mountains, rivers, capes, and bays were often misrepresented, and it also failed to give details because of its very small scale, which was about 1:1300000. In comparison with other maps published before, however, its accuracy was remarkable, and naturally it was so warmly received by the public that it went through several editions.

(*D*) *Other Sciences*

When Europeans had once opened communication with this country, information on the geography, industry, form of government, and customs of Europe was of great importance and interest to the Japanese; and every time Europeans arrived some fragmentary information was circulated. Then, of course, Japanese envoys like Tsunenaga Hasekura must have come back with a great deal of such information. However, the Japanese had no definite knowledge of Europe worth mentioning before the end of the 17th century. An Italian by the name of Sirotti had come to Yakushima in Kyūshū in 1708. The Shogunate sent him to Yedo for imprisonment and ordered Hakuseki Arai (1657–1725) to inquire from him regarding the geography of Europe and many other things. Consulting the world map which had been stored in the Shogunate library, Arai recorded what Sirotti had to tell; and also taking into consideration what he had heard from the

Dutch captains who had come to Yedo, he compiled in 1713 a geography of the world, entitled "Sairan-igen." Then on the basis of the conversation with Sirotti, he also wrote "Seiyōkibun" (Occidental Informaton) in 3 volumes. Since these books record only what the prisoner answered to his questions on several occasions through the very imperfect translation of interpreters, the contents are naturally coarse and meagre. Yet they are the first books on the geography and customs of Europe. Following the lifting of the ban on foreign books, Chinese translations of European geographies and various maps came over to Japan. Modifying the contents of these works with what they had heard, some others wrote books on the state of things in Europe. No book had given the real truth as yet. It was not until the Japanese scholars had facilities for the study of the Dutch language, and opportunity for themselves to read the original works which had been brought over from Europe, that they wrote books containing comparatively detailed information of European geography and other subjects. Saisuke Yamamura (1770–1807), encouraged by reading the Sairan-igen, and deliberately planning an exhaustive study of world geography, set about studying Dutch first, and, after reading various foreign works, wrote a revised "Sairan-igen" in 1802, ninety years after the publication of the original by Arai.

It may be well to say a few words here on the development of the knowledge of the Dutch language. Formerly the interpreters at Nagasaki were expected to act as interpreters only through hearing and memorizing Dutch, but not allowed to read Dutch books. In 1745, however, this ban was removed. About this time Bunzo Aoki (1693–1769), a samurai of the Shogunate, noted for his cultivation of potatoes, coming to Nagasaki, tried to study Dutch, but did not attain any great success. A physician of the clan of Nakatsu, named Ryōtaku Maeno (1723–1803), intending to study Dutch, though more than forty years of age, took some lessons under Aoki, and studied at Nagasaki for a hundred days. Still he did not find his knowledge of the language fit for practical use. With the help of a crude dictionary which he had taken to Yedo with him, he educated himself in the language to some extent. At that time there was a physician of the clan of Obama named Gempaku Sugita (1733–1817). At his request his feudal lord bought a copy of a Dutch book entitled Tafel Anatomia, the contents of which he could not read, but could only roughly guess from its illustrations. When, in the spring of 1771, a post-mortem examination of an executed prisoner took place at the Kotsukappara execution-ground in Yedo, Sugita went to see it with Maeno and another man. They found the internal organs to be exactly like the illustrations in the Dutch book. Thereupon, they were determined to translate this work in spite of all difficulties and to contribute to the study of medicine in Japan. On the very next day, they gathered comrades together and set about the task of translation. Since Maeno was the only one in the group who had some knowledge of the Dutch language, the work was immensely difficult, and it was four years before they at length published the Japanese translation, "Kaitai-shinsho" (A New Book on Anatomy) in the fall of 1774. It goes without saying that this book gave a tremendous stimulus to the medical world. As a result of these labors, Maeno, Sugino, and others considerably increased their knowledge of Dutch, which aroused other students of medicine to the absolute necessity of learning the foreign language. Gentaku Ōtsuki (1757–1827) contributed more than any other to the advancement of the study of Dutch. This man studied the language first under Sugita and Maeno, and then went to Nagasaki for further study. He not only translated various Dutch books into Japanese, but also wrote a book entitled "Rangaku-Kaitei" (A Dutch Primer) as a guide for people who wished to study the language. Thereupon, a large number of people were inspired to study it. Able men like Sampaku Inamura, Saisuke Yamamura, Sōkichi Hashimoto, and Genshin Yasuoka became pupils of Ōtsuki and the study of the Dutch language received a sudden impetus towards the end of the 18th century.

Again, an inquiry into the introduction of machines of European style into Japan shows that articles of daily use such as clocks had already been brought over in the 16th century. The clocks had no hair springs, and were controlled by horizontal levers. Some people during the early part of the Tokugawa period were already clever enough to make imitations of such clocks. Telescopes were also introduced from an early date, and some were even made in Japan at the beginning of the 18th century. The Shogun Yoshimune ordered the making of new instruments with which to observe the heavenly bodies, and also demanded a constant gauge of the amount of rainfall. Taking advantage of this inclination of the Shogun, the Dutch merchants must have brought in newly invented machines. What kinds of machines these were that came over at this period is not clear. In the latter half of the 18th century, it is apparent that electric machines, barometers, thermometers, cameras, magic lanterns, etc., were introduced. Gennai Hiraga (1733–1779) who was naturally fond of natural science went to Nagasaki to associate with foreigners and gain new knowledge. Having seen an electric machine for the first time in 1770, he made an imitation of it, and this is still preserved by his descendants. Thermometers were made in Japan towards the end of the 18th century. However, these were imitations and simply regarded as curious articles. Towards the end of the century, Ryōtaku Maeno translated part of a work on dynamics. Among the works by Sōkichi Hashimoto, some relate to certain physical problems. But the knowledge of physics of the Japanese at the time was decidedly infantile and fragmentary.

It was not until twenty or thirty years had elapsed that books treating the subject somewhat systematically began to appear.

We should here mention Kōkan Shiba (1747–1818), the first Japanese to introduce western arts into our land. Shiba at first studied the genre painting under Harunobu Suzuki, and called himself Harunobu II., succeeding to his master's name. As he further studied under Bunchō Tani, his art still progressed. When he saw a western painting, he became anxious to learn the art, which he finally mastered after assiduously studying under a Dutchman in Nagasaki. While in Nagasaki, he wished also to learn the method of copper-plate etching, and at last succeeded in this. His copper-plate prints came out in 1783. The first copper-plate prints that were made in Japan date as far back as 1591, but the method had been long lost. Through his efforts, Shiba revived it. Going to Nagasaki again in 1789, he further studied oil-painting and copper-plate printing. Later on he published a number of delicate copper-plate prints, such as the plate of the celestial sphere and that of the terrestrial globe, etc., through which he did much directly and indirectly to further the advancement of science in our country.

CHAPTER II

REMARKABLE DEVELOPMENT IN THE STUDY OF WESTERN ASTRONOMY IN JAPAN

As we have seen, the Japanese calendar that Shunkai Shibukawa adopted in 1684 was corrected in 1754 and again in 1769. This calendar system was worked out on the basis of the Shou-shi Calendar and was necessarily not very accurate; therefore, as time elapsed, errors so accumulated that towards the end of the 18th century, the calendar needed some fundamental changes. Hereupon, the Shogunate first ordered its astronomers, Tokufū Yamaji, Hidemasu Yoshida, and others to construct a tentative calendar according to the method introduced in the "Ch'ung-chên-li-shu." But at that time the Sequel of the "Li-hsiang-k'ao-ch'êng" (compiled by J. Kögler, A. Pereyra, and other missionaries stationed in China) which introduced Kepler's laws of elliptic orbits had already come into Japan, and the theory of the Ch'ung-chên-li-shu was considered to be antiquated. Somehow the high officials of the Shogunate suddenly became ambitious and again ordered a new calendar to be compiled according to Kepler's new laws described in the Sequel of the Li-hsiang-k'ao-ch'êng. Contrary to the expectation of the officials, the astronomers of the Shogunate had not mastered the contents of the Sequel and were not equal to the task of constructing a calendar on the basis of this theory. So the officials began to raise astronomers among the people who could assist in their calendar-construction. They invited Gōryū Asada who was most popular in Ōsaka as a master of astronomy. Asada excused himself for a certain reason, but recommended two of his best pupils—Yoshitoki Takahashi and Shigetomi Hazama.

who, entering the Astronomical Bureau of the Shogunate, constructed a new calendar in 1797, which was based on the Sequel of the Li-hsiang-k'ao-ch'êng and somewhat modified by what Asada had designed. This was the "Kwansei Calendar." Afterwards Takahashi obtained a Dutch translation of the Astronomia by Lalande, which he studied assiduously. On his passing, Shigetomi Hazama, and two sons of Takahashi—namely, Kageyasu Takahashi and Kagesuke Shibukawa—carried out their master's dying wishes and finally succeeded in mastering and digesting the work. Consequently the study of western astronomy in Japan achieved remarkable progress before that of any other western science. Asada should be called the very pioneer of the study of western astronomy in Japan; and Takahashi and Hazama considerably developed and extended his work. Takahashi's two sons then gave it a finishing touch. The efforts of these men should not be overlooked in recording the development of the study of astronomy in Japan. Besides, Takahashi, Hazama, and Takahashi's two sons were on intimate terms with Tadataka Inō, hero of the present work. We may properly discuss their achievements here.

1 Gōryū Asada

Asada's true name was Yasuaki, but he is better known by his pen-name, Gōryū. He was born in 1734, the fourth son of Yasumasa Ayabe, a samurai of the clan of Kidzuki in Bungo Province. Fondness for astronomy marked his childhood; the names of the chief planets and stars once mentioned to him never were forgotten. At the age of about six, it is said, he used to mark upon the verandah how far the sun-light came under the eaves throughout the year, and astonished his family by discovering that the sun gradually travels north from south after the winter solstice through spring and part of summer, and returns south after the summer solstice. At about the same age, he is also said to have found out for himself that the fixed stars travel westward in relation to the sun. This indicates that Asada, even in childhood, was quite different from ordinary children, and was an astronomical prodigy. As he grew, he undertook to study medicine and astronomy. Tradition says that he never sought a teacher, but educated himself, working ceaselessly and arduously till he mastered all the subjects. It must be quite true that no teacher was responsible for Asada's mastery of the science of astronomy. Neither can it be denied that he had access to those silent masters and companions—valuable books of reference. As Japanese books written on the basis of the Shou-shi Calendar were abundant and popular in his youthful days, his first step in the study of astronomy must have been to read some of these works. Besides, it was now much easier to obtain such books as the Li-suan-ch'üan-shu and the Ch'ung-chên-li-shu which treated of western theories; and, that he should have acquainted himself with some, if not all, of the works of this kind and informed himself regarding the western theories is to be inferred from the contents of his books. What is to be noted here especially is the fact that Asada placed great weight upon practical observation, and every time he came across a new theory he determined its value by the results of his own observation. Subsequently, he was constantly engaged in making observations of the solar and the lunar eclipses and of the motion of the heavenly bodies. The oldest existing record of his observations is that of a lunar eclipse in 1757 and it shows that he was a thoroughly experienced observer at the age of about twenty-three. It is only natural that he should have paid attention to the improvement of astronomical instruments as a natural result of valuing observations. It is evident from his letters to his friends that he contemplated this even in his early days. When the state calendar, because of its defectiveness, miscalculated a solar eclipse in 1763, he had already pointed out, the year previous, the error in the official calendar and showed the result of his own calculation to his friends. As the day arrived, the eclipse exactly coincided with his calculation. It is apparent that his capacity then as a student of astronomy was far superior to that of the official astronomers of the Shogunate.

Asada was dependent upon his father until 1767. Though he had no income of his own, he was never molested by official duty; he was free to give himself up to the study of astronomy and medicine. However, he was appointed in that year a physician of the clan, and as he was ever on the move, accompanying his feudal lord to Yedo or Ōsaka, he now found it impossible to pursue his favourite study of astronomy. So he repeatedly implored his lord to excuse him from service, but in vain. Making up his mind at last, he deserted his clan and went to Ōsaka, where he arduously resumed his study of astronomy, making his living by practising medicine. The various records fail to agree as to the year of his coming to Ōsaka, but since there is a record of his observation of a lunar eclipse in 1769, he must have been there by that time. It was during his residence in Ōsaka after deserting his clan that he gave up his family name Ayabe and assumed a new one—Asada.

On coming to Ōsaka, Asada was more than ever concerned with making observations and improving instruments. Polishing his own lenses, he constructed telescopes. He also designed a new type of telescope in which the image by the lens could be produced on a suitable place by means of a reflector. By means of such instruments, free pendulums, and quadrants, he applied himself to making observations day and night and, after strenuous efforts, constructed a calendar called "Jichū-reki." Of the books he wrote, only a few are now existing. One of them entitled

"Jikkenroku" (Records Based on Observations) gives the essentials of his calendar: taking the winter solstice of 1781 as the origin of time and Ōsaka as the standard station, he gives the fundamental constants and his method of calculations for figuring out the positions of the sun and the moon; and also his method of calculating solar and lunar eclipses. This book has an endorsement by Asada in 1786, but the fundamental constants adopted in the work were revised twice—in the spring of 1793 and in the winter of 1794. In marking the motions of the sun and the moon, he adopted in this work the method described in the first volume of the Li-hsiang-k'ao-ch'êng—that is, the motions of the sun and the moon are designed to be composed of a series of uniform circular motions by superposing a number of circles in succession, such that the centre of one circle revolves along the circumference of the other. The constants employed for calculation were mostly new ones which he had worked out on the basis of the results of his own observations and which he had tested by the solar and lunar eclipses in the past. However, some of the constants he adopted in 1786 are identical with those given in the first volume of the Li-hsiang-k'ao-ch'êng, and the others are likewise more or less similar to the figures in the same work. Hence it may be inferred that until 1786 when he wrote the Jikkenroku, Asada had been studying western astronomy chiefly by means of the first two volumes of the Li-hsiang-k'ao-ch'êng; and he had been expatiating upon the work in the light of his own observations and speculation. The most remarkable thing is that in the spring of 1793 he made considerable correction on the constant in giving the distance between the sun and the earth, and also corrected other constants affected by it. That the corrected constant in indicating the distance from the sun is almost identical with that given in the Sequel of the Li-hsiang-k'ao-ch'êng shows that Asada had access to the Sequel about this time, when he came across the theory of the elliptic orbit, and became aware of the gross error of the constant for the distance of the sun given in the first volume. It seems that Asada retained the old theory concerning the mode of the motions of the sun and the moon while he worked radical corrections on the important constants evidently for the purpose of testing their adaptability. Ever since Asada obtained the Sequel of the Li-hsiang-k'ao-ch'êng through the efforts of Shigetomi Hazama, one of his pupils, he and his best pupils occupied themselves in the study of the theory of the elliptic orbit, and even before the official astronomers of the Shogunate could master the new theory, they had already digested it so thoroughly that they could apply it freely to the practical problems.

Thus Asada's method of calendar-construction was based upon the first two volumes of Li-hsiang-k'ao-ch'êng, but the chief characteristic of his work, distinct from that of the West, consisted in the fact that the most of the constants employed were not absolutely constant, but set as changeable with the lapse of time—that is, he adopted a system of periodic correction. The idea of such correction did not originate with him. In the T'ung-t'ien Calendar of the Tsung dynasty, such a system of correction had been employed for the length of the tropical year, and the Shou-shi Calendar had also followed it. The variation of the inclination of the ecliptic also had been known for a long time. However, Asada's system of correction was to be applied not only to these, but also to the lengths of the synodical month, the nodical month, and the anomalistic month, and other constants; every one of which was set as changeable with the lapse of time, but practically treated as unchangeable for every ten years, after which all the constants were to be revised according to the system, so that they might be available the next ten years. Though his original manuscript of this theory has been lost, a copy of his system of correction which Takahashi and Hazama, his favourite pupils, submitted to the calendar officials of Kyōto in 1797 when the calendar was revised, and another copy that Takahashi enlarged and revised in later years have been preserved to this day, and these give us an idea of the substance of the theory. As one looks into these existing works, it is evident that Asada's deductive method in designing the system of correction lacked not a little in exactness. Moreover, in comparison with the figures adopted by the astronomers of recent years, some of his figures differ so widely that they would have been far simpler and more accurate without his correction. However, it is to his credit that the idea that only the tropical year and the inclination of the ecliptic were subject to change did not satisfy him, and that he, assuming most of the constants which had always been accepted as absolutely constant to be functions of time, worked out a system of correction for them.

While engaged in the study of astronomy for years, it is presumed that Asada invented many things besides the system of correction, though there remain few authentic records. What is the most remarkable of all, is the fact that Asada had been aware of the relation between the distances of the planets from the sun and the periods of their revolution—that is Kepler's third law—before this was introduced. The Sequel of the Li-hsiang-k'ao-ch'êng treated of the elliptic orbits of the sun and the moon, but not of planets, so naturally it did not touch this third law. The law was first made known in Japan in 1803, after his death, as Takahashi and others obtained the Astronomia written by Lalande. On coming across this third law in the work, Takahashi wrote, "This is what Mr. Asada, my late master, already conceived. What a wonderful coincidence that the same theory should have been conceived in the Orient and in the Occident!" The record of this impression is still preserved. Yet in what year and with what motive Asada lighted upon this law is not clear. As a result of his observations for years, he is entitled to the honour of being the first Japanese to observe many astronomical phenomena. Among other things, he determined the period of the sun's rotation by observing the sun-spots, studied the aspects of the moon by carefully observing her surface, measured the periods of revolution of the four satellites of Jupiter, and discovered that the rings of Saturn changed their aspects with the lapse of time.

In this manner, Asada, after settling down in Ōsaka, was busily engaged for twenty years in making observations, and in expounding western astronomy which he had modified by his own theories, and also in educating his pupils, among whom he boasted of Yoshitoki Takahashi, Shigetomi Hazama—two great geniuses—and many other able scholars. Consequently Asada's popularity grew day by day until he came to reign supreme in the Kwansai region as the leader of the school known by his name, and Ōsaka became, as it were, the nursery of scientists. For this reason, when the Shogunate proposed to revise the Japanese calendar by the new theories of western astronomy and found out that its own official astronomers were inadequate for the purpose, it attempted to appoint Asada as the reviser of the calendar. Now, when Asada deserted his clan and hid himself in Ōsaka, his feudal lord to whom his whereabouts was known was generous enough not to charge him with the crime of desertion; and, besides, he was even permitted to communicate with his relatives at home through the generosity of his master who was eager to see him succeed in his pursuit of learning. This made it possible for Asada to receive a gift of some money each year from his eldest brother and to indulge in the study of astronomy without suffering from extreme destitution. For this Asada was so grateful to his former master that when he became a well-known scholar and was offered a high position by other feudal lords, he always refused, saying that he could not turn his back on his former lord. Even when he was invited by the Shogunate, he still refused the offer, deeming it his duty to remain loyal to his former lord. Instead, he recommended as his substitutes his great pupils, Takahashi and Hazama.

In studying medicine, as in the study of astronomy, Asada collected and read its literature extensively, and gathered up the very best that was to be found. He paid special attention to the relation between the efficacy of medicinal substances and the treatment of diseases. He also dissected dogs and cats, through which he became acquainted with the conditions of the internal organs, and went through a long course of experiments. From the fact that Asada expounded western theories in his study of astronomy, some people presume that his medical art was copied from some Dutch school of medicine. This is entirely wrong. According to a reliable record, Asada chiefly adopted the medical art that had existed in Japan since the earliest days and, as a result of his efforts, discovered sixteen efficacious prescriptions which he made it a rule to give. At the time that Asada studied astronomy and medicine at home, there were a few Japanese physicians, it is true, who taught something of a very elementary surgery in western style, but there was no proper literature for students. As for the western style of internal treatment, none knew about it, and naturally there was no book written on it. It is but natural that Asada should have had no access to medical science of western style, because he had to study medicine chiefly through reading books in a remote place. In spite of this his skill in treating patients was remarkable. While he was a court physician, his feudal lord was once seized with a sudden illness and was suffering from acute pain. The best the physicians could do was of no avail. Then Asada requested to be put in charge of the case, and in spite of the others' opposition, administered a medicine of his own preparation to the patient, whose pain was immediately relieved. However, this service of Asada's, it seems, proved a cause of jealousy on the part of the unworthy physicians and also a remote cause of his deserting the clan.

In Ōsaka he practised medicine, but it was only a means by which he might make his living. All his energy was devoted to the study of astronomy. Now that his research was temporarily completed and, in addition, Takahashi and Hazama on whom his mantle had fallen accomplished the task of revising the state calendar in 1797 and Asada himself was rewarded by the Shogunate, he could say, "I have found men who will develop my astronomy. I have hence to devote myself to the study of medicine." It seems that while Asada was absorbed in the study of astronomy, the Japanese physicians were gradually stimulated by western medical art; especially after the "Kaitai-shinsho" (A New Book on Anatomy) was published, a large number of students aspired to study medical science in European style. The times were mature in 1793 for the publication of a Japanese translation of a Dutch treatise on internal treatment. All this must have induced Asada to study the new school of medicine thoroughly. But he had not succeeded in this before his health broke down about the beginning of 1798, and he suffered from increasing infirmities of age, of which he at last died in the summer of 1799 at the age of 65. His services in the development of astronomy in our country remained unrecognized for a long time; but of late years people have become aware of his contribution, for which the Emperor conferred upon him a posthumous title of Junior Grade of the Fourth Rank.

2 Yoshitoki Takahashi and His Two Sons

That the study of western astronomy expounded by Asada attained such remarkable progress at the beginning of the 19th century was really due to the efforts of his greatest disciple, Yoshitoki Sakuzaemon Takahashi (1764–1804). Takahashi is generally known by his pen-name Tōkō. Born the son of a samurai of low rank in Ōsaka, he succeeded to his father's position of *dōshin*, or police constable, at the age of fourteen. He was fond of mathematics and astronomy from childhood, and studied Japanese mathematics with a certain Nōichi Matsuoka and astronomy with Asada. It is not clear in what year he entered Asada's school, but we can safely assume that the date was before 1790. He was extremely poor, but had the good fortune to be helped by a clever wife. Since persimmon-raising was a supplementary source of his income, when the fruit was ripening, he had to expend much time and labour to guard it against thieves, and this interrupted his study of astronomy. Therefore, while he was away once, his wife had every one of the persimmon-trees felled, thus doing away with the source of his disturbance, though it meant some sacrifice on her part in house-keeping. This anecdote will serve to suggest the assistance his wife gave him in her sympathy with his studies. It is a great pity that she passed away without witnessing her husband's distinction.

The benefits gained by Takahashi at Asada's school through the latter's direct guidance must have been immense, but the greatest advantage was that of his having an opportunity to make the acquaintance of Shigetomi Hazama and, through his aid, to read and master the Sequel of the Li-hsiang-k'ao-ch'êng, then the rarest book among the common people. Instead of advocating a method to represent the actual motion of a heavenly body by composing several uniform circular motions as customarily employed in the first and second volumes of the Li-hsiang-k'ao-ch'êng and every other astronomy previous to this, the Sequel expounded a method to represent it by a non-uniform motion revolving on an elliptic orbit under a certain law. Thus the motions of the sun and the moon were unified under a simple law, and the principles of calendaric computations were made extremely lucid. However, to compute the positions of the sun and the moon according to this new law, a thorough knowledge of the mathematical properties of an ellipse was necessary and the method of calculation involved a far greater difficulty than that used in the old style. By this time Japanese mathematics had already made remarkable progress, the students of western astronomy generally having little difficulty in the matter of mathematics. But in connection with an ellipse, as no one had yet studied its properties; the readers of the Sequel of the Li-hsiang-k'ao-ch'êng rather keenly felt a mathematical difficulty, and especially for a student with no special scientific turn of mind, it was no small task to master the theories in the volume and to put them to practical use. This was the reason why the volume, which had come into the official Astronomical Bureau long before the private scholars were aware of its existence, had been left unutilized, but foolishly kept under lock and key. Once in the possession of Hazama, this book became an object of devoted study on the part of all Asada's advanced students, and even the master himself. Their efforts enabled them to masticate and digest the principles contained in the volume, and subsequently to effect a reformation in the astronomy of the old school. And of all the group, Takahashi was the one most thoroughly acquainted with the new theories and capable of most readily computing new combinations and putting them freely to practical use. As he read this work, he made sacrifices of food and clothing in constructing various astronomical instruments of a new style and still pushed on in his investigations. He enlarged and revised Asada's system of correction, which he introduced into the theories of the Sequel of the Li-hsiang-k'ao-ch'êng to develop a new system. His real abilities in astronomy about 1795 immensely surpassed those of his fellow-students, and even those of Asada, his own master. He was now the outstanding scholar of astronomy throughout the country.

It was in the spring of 1795 that Takahashi in place of Asada was summoned to Yedo to take charge of revising the calendar. He, with his comrade, Hazama, was engaged in making astronomical observations and in planning a new calendar as a preliminary to their task, and was appointed an official astronomer of the Shogunate at the end of the same year. The following year he, accompanied by other official astronomers, went to Kyōto according to precedent, where they made observations to prove the accuracy of their new calendar and discussed the formal requirements of a new calendar with the calendar officials of Kyōto. They reported to the Throne about their new calendar which the Emperor graciously named the "Kwansei Calendar," Kwansei being the name of that era.

The Kwansei Calendar was modelled after the Sequel of the Li-hsiang-k'ao-ch'êng and modified by a system of correction. Takahashi and Hazama were solely responsible for the design and the other official astronomers appeared to be sitting for formality, but as it was still in the days of the old Shogunate régime when social standing and precedents were respected, the senior officials, it seems, were inclined to oppress Takahashi and others to such an extent that the latter could not persist in their contention as they desired. As to the motion of the planets, no new method was mentioned in the Sequel of the Li-hsiang-k'ao-ch'êng. Neither were the data of observations made in Japan for newly determining the constants necessary to compute the planetary position by the theory of an elliptic orbit sufficient in quantity. For this, they temporarily retained the old system in calculating the positions of the planets. As soon as the task of calendar-construction was over, therefore, Takahashi, planning to design a system of computing the position of the planets, was engaged in making observations, first of Mars and Venus, and then of Saturn, Jupiter, and Mercury. On the occasion of a lunar eclipse in 1798, Takahashi discovered an error of about 7·5 minutes between the time of his observation and the calculated time. ' This convinced him that the computations concerning the lunar motion given in the Sequel of the Li-hsiang-k'ao-ch'êng were not very reliable, either; and he felt the need of constant observations which he now tried to carry on. He was not merely clever in making observations, but he also did his best to prevent practically as well as theoretically the occurrence of errors, giving careful consideration to his adjustment of instruments and the causes of errors; and by this reason, the accuracy of the results of his observations was remarkably superior to anything done by ordinary observers with the same instruments.

Since Takahashi became an official astronomer of the Shogunate he was given the privilege of getting access to all astronomical literature; and he revised the values of his own by the constants which he found scattered in the Dutch books treasured by some feudal lords, and simplified the theories contained in the Sequel of the Li-hsiang-k'ao-ch'êng. He also improved the method of computing eclipses and other phenomena by his own design until the books and papers he wrote numbered several dozen. The contents of those books and papers all sufficiently prove his clear-headedness, superior to that of ordinary astronomers.

While pursuing his own studies, Takahashi at the same time taught his pupils and helped Tadataka Inō undertake the grand enterprise of surveying the whole of Japan. To begin with, in his astronomical computations, Takahashi often felt the necessity of determining the longitudes and latitudes of various important points in Japan as well as the dimension of the earth. As to the earth's size, foreign books varied, some allotting 60 miles to a degree of the meridian and others 15 miles. There was no knowing what a mile was in Japanese measurement. What the Japanese had held till then was also varying; the length of a degree was held to be 30 *ri*, 32 *ri* or even 25 *ri*, none of which was a result of accurate measurements, and was not to be relied upon. The latitudes of some important points of the country had been determined through observations, but as to their longitudes, almost no observation had been made. This greatly embarrassed Takahashi in his astronomical computations. At this time Inō was contemplating a geodetic career. Moreover, Russia was eagerly watching for a chance to encroach upon our northern frontier so that the Shogunate keenly felt the necessity of defining the territory of Yezo from the standpoint of defence. This fact led Takahashi to believe that if he should let Inō apply for permission to make an accurate map including Yedo and Yezo, the government would readily grant this, and incidentally the problem as to the length of a degree would be solved. So he discussed the matter with Inō, who was persuaded to make up his mind to carry out the survey at his own expense. Takahashi then broached the subject with the Shogunate and obtained the desired permission after considerable negotiations. The way was thus cleared for Inō to carry out a survey of all Japan. In forwarding the opportunity of this geodetic enterprise, Takahashi's intercession and painstaking efforts were inestimable. But for his master's efforts, Inō would not have had the chance to apply his abilities to land-surveying.

As a result of a survey in 1801, Inō obtained 28·2 *ri* as the length of a degree of the meridian, but Takahashi had some doubt as to this figure. However, as Takahashi had an opportunity for the first time to read the Astronomia written by Lalande in the spring of 1803, he was convinced that the length of a degree detemined after careful measurements coincided with Inō's result, and that all the theories given in the work were wonderfully accurate, out of all comparison with those in the Sequel of the Li-hsiang-k'ao-ch'êng, the only thesaurus accessible to Japanese astronomical students up to that time. Thereupon, he bent his whole energy to the mastery of this work and the revolutionizing of the astronomical theories. The volume Takahashi was allowed to read belonged to a certain private person and had to be returned in a fortnight. So he petitioned the authorities to purchase it, and in the late summer of the same year, the authorities gave consent and purchased the volume for Takahashi and other official astronomers to read and keep it permanently. Though Takahashi was now engaged in expounding this Dutch astronomy, he experienced considerable hardships because of his lack of knowledge of the Dutch language. His whole acquaintance with it was the alphabet, some nouns, and the use of a dictionary. However, as Takahashi had among his acquaintances Gentaku Ōtsuki and other veteran scholars of the Dutch language, and also some young students whose assistance he could readily enlist, it is possible that he was directly and indirectly assisted by these Dutch scholars in studying the foreign work. Even in that case it goes without saying that those who knew ordinary Dutch must have come across abstruse terms and often have failed to give the exact meaning to Takahashi. With his unusual zeal and profound scholarship, Takahashi in spite of his shortcomings in the matter of the language came to comprehend the chief arguments in this astronomical work and compiled more than a dozen volumes entitled "Lalande-rekisho-kwanken" (A Review of Lalande's Astronomia). This consisted of a rough translation of some passages of the book considered especially important and his own comments upon them. This was a task which Takahashi accomplished even within six months of his receipt of the Dutch book. How assiduously he worked can be inferred from this.

Takahashi had been suffering from ill-health for a long time, and it would seem that from as early as 1800 he had sustained his weak constitution chiefly by drugs. Since obtaining the Astronomia he worked day and night till his health was so affected that he had to take to his bed. Even then he did not put down his book, but still continued the work of translation. His illness advanced and in February, 1804, without having fully accomplished his purpose, the great astronomer breathed his last in the official residence of the Astronomical Bureau in Yedo. Alas! a genius of unrivalled scholarship and technical skill—and a man who, in an evil age when scholars usually attempted to exclude one another and to secure their own advancement at the expense of others, had stood firmly as adamant and had become the forerunner of the new science—what a pity that such a man should have died when he was just ready to pave the way for his followers to advance and when so full of promise of greater achievement! If Heaven had given this man a few more years, a great deal more would have been accomplished in the advancement of the study of astronomy in Japan, and Inō's enterprise of surveying and map-construction would have been considerably more exact and accurate. What a great loss to the world of science that such a man should have died at the early age of forty! When he became critically ill, he asked his bosom friend Hazama to help his heir translate the Astronomia and thus accomplish his own purpose. This task of translation was continued by Hazama and Takahashi's two sons, and was completed in 1836.

Takahashi had two sons. The elder was Kageyasu (1785–1829) who named himself Sakuzaemon Takahashi after the death of his father. The younger was Kagesuke (1787–1856) who becoming heir to a descendant of Shunkai Shibukawa named himself Sukezaemon Shibukawa. The elder was a master not only of astronomical computations, but of Dutch and Manchu. After his father's death, Kageyasu succeeded to his father's position of official astronomer of the Shogunate, and also officially supervised Inō's enterprise of surveying. He was always deeply interested in the state of things in foreign countries, and learned much through extensive reading, being gifted with linguistic talent. He translated not only the Astronomia—the task left by his father—but many other western books as well. He also wrote several books on the Manchu language. From Rinzō Mamiya's report, he wrote an informative book on the topography and the state of things of Saghalien and the land beyond the northern frontier; and by order of the Shogunate, he then compiled a map of the world. The copper-plate engraving of the map was completed in 1810, and each hemisphere on a stereographic projection, about 1 m. in diameter, was in much more detail than any previous one. About this time he also made a copper-plate engraving of a sketch-map of the Japanese frontiers, on which the Loo-choo Islands, Kamtchatka, and the intermediate islands, and the coasts of China, Korea, and Manchuria, were plotted by a conic projection. This proved a handy map for a brief glance at the topography of the Far East.

A translation office was established within the Astronomical Bureau in 1811. Most of those proficient in the Dutch language were then transferred to it, and became engaged in translating western books into Japanese. It appeared as if the researches of western civilization were centered within the Astronomical Bureau at Yedo. By this time Kageyasu had more and more improved his knowledge of the foreign languages and come to enjoy the deep confidence of the Shogunate. He was now promoted head of the official astronomers and was at the same time commissioner of the libraries of the Shogunate. In the spring of 1826, a German physician named Ph. F. v. Siebold, came to Yedo accompanied by a Dutch captain. Kageyasu frequently interviewed him at his lodging and, stealing a glance at some books written by a Russian and a new map of the Dutch territories which Siebold possessed, he thought that a translation of such books would prove of great service to the country. Complying with Siebold's wish, Kageyasu gave to the foreigner a small survey-map of all Japan by Inō which he had had copied, and also lent him some books, in return for which Siebold gave the coveted books to Kageyasu. The fact that Kageyasu had given a survey-map to a foreigner, violating the law, was revealed in the winter of 1828, and he was arrested and imprisoned. Before his sentence was given, he died in the prison after several months' imprisonment.

Kagesuke, his younger brother, was of a gentle disposition and also fond of learning, and while a young man, he accompanied Inō in his survey of the country. Though the Shibukawas had been in charge of calendars since the days of Shunkai and the most ancient family among the official astronomers, his descendants had lacked in the knowledge of astronomy, and lagged behind in the study and business of calendar-construction. When Kagesuke succeeded to the family, he soon distinguished himself. He was a keen student of western astronomy, and was engaged in the translation and expounding of the Astronomia with his brother. After his brother's imprisonment, he continued the task single-handed and completing it in 1836, he entitled it "Shinkō-rekisho" (Astronomy of a New Design). In 1841, he was ordered to construct a new calendar according to the Shinkō-rekisho; and in the following year, he reported to the Throne upon a new calendar, which the Emperor named the "Tempō Calendar," adopting the name of the era. This was the calendar enforced in Japan until 1872, when the solar calendar was adopted. The

Shibukawa family owned a large amount of astronomical literature from the oldest times. And as Kagesuke was naturally conscientious, he never slighted even the most trifling document, but took up, studied, and readjusted them carefully so that a great number of books appeared as a result of this. It is mostly due to these documents that we are able to-day to examine our old calendar systems scientifically. Unlike his short-lived father or brother, he enjoyed a long life, dying at the age of sixty-nine.

3 Shigetomi Hazama

A man who, standing side by side with Takahashi, contributed much towards the advancement of the new science of astronomy was Shigetomi Gorōbei Hazama (1756–1816). Like Takahashi, he was born in Ōsaka, but in a family of considerable wealth, which had conducted a pawnshop for generations. This man was likewise fond of mathematics and astronomy from childhood, and was a genius of technical skill. As he grew older, he commenced to read astronomical literature from its earliest times, and realizing the greater accuracy of western astronomy, was educating himself with eager study, when he heard of Asada's reputation as a master of western astronomy and went to study under him. This was at about the same time as Takahashi did. Those expounding western astronomy at that time all depended upon the Ch'ung-chên-li-shu and the first two volumes of the Li-hsiang-k'ao-ch'êng, whereas the Sequel of the latter work introducing the new laws of the elliptic orbits was never made use of, vainly stored away in the official quarters and in the secret libraries of the nobles. Hazama deeply regretting this, and trying by every means in his power, finally succeeded in obtaining a chance for the eager unofficial students to peruse it. The book which Hazama, while a pupil of Asada's studied anxiously with his master and fellow-scholars was this Sequel, and the result of this research alone enabled him to become, together with Takahashi, the ruling spirit in compiling a new calendar in later years.

Hazama, while studying the principles of astronomy, worked hard at the improvement of astronomical instruments. He trained new mechanicians, and had various instruments made, which marked a new epoch. Asada's greatness may be accounted for partly by his mastery and digestion of western astronomy and partly by his regard for observation and his constant application of this with improved instruments. By his accurate observations he tested different theories in order to adopt the very best he could find. The instruments Asada employed, though somewhat superior to those used among the unofficial astronomical students, were by no means exact. The more exact the astronomical theories became, the more precise were the instruments acutely needed. It is true that books minutely illustrating the method of constructing a quadrant and other astronomical instruments in the new style had been brought from China to Japan, and such an instrument as a pendulum-clock, though not described in those books, might have been seen, or a diagrammatic illustration of it at least, by the scholars. Therefore, the greatest difficulty that Asada and others encountered in constructing their new exact instruments was, not how to make them, but how to get mechanicians capable of constructing them and how to meet the cost. Hazama solved this problem of mechanicians and expenses, making good use of his wealth. The transit-instrument, the quadrant, the pendulum-clock, etc. which were newly constructed in Kyōto for the compilation of the new calendar, were all done by Tōzaburō Toda and others, the mechanicians Hazama had trained. He had many instruments constructed by his own design:—such as a simple eclipse-meter, an ellipsograph, a barometer, etc. Invited by the

Shogunate to compile a new calendar with Takahashi, he went up to Yedo in 1795, and having discharged his duties, returned to Ōsaka in 1798. While in Yedo, devoting himself to this task, he sowed in various quarters seed of the development of astronomical study. For instance, during this time, he trained Yagorō Ōno, the official mechanician to the Astronomical Bureau at Yedo, and Yasaburō, his son, to construct new astronomical instruments; he answered the questions of Inō, one of Takahashi's pupils, did much toward enlightening the latter, used his influence in purchasing astronomical books and a pendulum-clock which had been constructed in Kyōto, and finally, personally supervised the construction of a quadrant for Inō.

Even after he returned to Ōsaka, by order of the Shogunate, he carried on a certain astronomical observation at his own house. In 1802, he was sent to Nagasaki to make precise observation of a solar eclipse for the determination of the longitude of that place. This plan was prevented entirely by rain. However, Hazama on his way from Ōsaka to Nagasaki and back surveyed his route by the traverse method, according to the plan which Inō employed in his survey of the north-eastern region and obtained enough materials to determine the longitude of Nagasaki geodetically. After a consultation with Takahashi, Hazama planned to make a survey of the Kwansai region as Inō had done in the north-east. He secured permission from the authorities, but just before he set out in 1803 for the survey of the Kinki section, he was taken ill. In addition to this he lost in a fire most of the measuring-instruments he had prepared; so he had to postpone the enterprise till the following year. At the beginning of that year, Takahashi died, requesting Hazama to continue his work. So Hazama went up to Yedo and became a member of the Astronomical Bureau again, and thus his plan to survey the Kwansai region was abandoned. Besides his work of translating Lalande's

Astronomia at his office, he assisted and coached Kageyasu Takahashi, and gave considerable support to Inō's survey work, which now extended over the Kwansai region.

For six years Hazama remained in Yedo, completing the translation of the various tables contained in the Astronomia. But now seeing all the enterprises of Takahashi well advanced and Kageyasu's genius brought into play, he asked leave and came back to Ōsaka in the spring of 1810. For a long time Hazama had taken a keen interest in the study of the evolution of Japanese weights and measures. Feeling the utmost necessity of a comparison of the European system with the Japanese, he had tried hard to settle the question finally. Taking the opportunity of coming back to Ōsaka, he made, with the Shogunate's permission, an extensive investigation of all the records and traditions of the ancient weights and measures relating to temples, Buddhas, temple-bells, grave-stones, Buddhist wares, Shinto wares, arms, stationeries, musical instruments, toys, etc., to say nothing of old rules, measures, and scales, which were preserved in the temples and shrines in Nara and Kyōto. If the report of this research were preserved to-day, it would provide many data on the evolution of our weights and measures, and it would also give a clear account of the origin of our system adopted by the Astronomical Bureau at Yedo. It is a great pity that it has been lost.

When Hazama left Yedo and returned to Ōsaka, he was supposed to be leaving only temporarily and was expected to come back again. He had been naturally of a weak constitution, and suffered frequent attacks of illness. This was hindrance to his full activity, and naturally never again was he able to proceed to Yedo. In the spring of 1816, Hazama died in his own house in Ōsaka at the age of sixty.

CHAPTER III

THE MONUMENTAL ENTERPRISE OF SURVEYING JAPAN

We have so far sketched the development of the sciences in Japan during the Tokugawa Shogunate, particularly towards 1800, and also the services of a few forerunners in advancing the new sciences. We have now arrived at the proper stage to describe the achievements of Tadataka Inō, the typical scientist of that age who, cleverly utilizing the sciences developed by these pioneers, accomplished the unrivalled task of land-surveying.

1 Inō's Earlier Life

Tadataka Inō, our greatest land-surveyor, was born February 11th, 1745 (according to the Gregorian Calendar, which we will employ always in giving dates)—the year in which the Shogun Yoshimune permitted the interpreters at Nagasaki to study foreign books. His father was Sadatsune Rizaemon Jimbō, the youngest son of a village headman of Ondzumi, Musa County, Kadzusa Province, but became the son-in-law and adopted son of Gorozaemon Ozeki of Ozeki, Yamabe County, the same province. He had two sons and one daughter. The youngest of these was Inō, and, therefore, he was to bear the name of Ozeki, but on the death of his mother in 1751, his father returned to the Jimbō family. He naturally assumed the family name of Jimbō, Sanjirō being his infant name.

Little is known about his childhood; but according to tradition, he lived with his mother's family till he was about ten years old, but upon her death he became a pitiable dependent, hardly having opportunity to learn reading, writing, and arithmetic, the common subjects of study. He spent most of his childhood taking care of the fishing-tackle at the Kujūkuri beach. When he came from his mother's home to his father who was still unable to make an independent living, but dependent upon the head of the family, though he had married again, the child found no happy home there, either. Until he succeeded to the Inō family at seventeen years of age, he was seldom with his father, but mostly lived with his relatives and friends scattered throughout the provinces of Hidachi, Kadzusa, and Shimōsa. However, this miserable life served to acquaint him with the ways of the world, to widen his experience, and to afford him an opportunity to study mathematics and medicine which had had an attraction for him from his childhood.

It is not clear under whom he studied mathematics and medicine. According to tradition, when about twelve years old, he begged a priest of a certain temple in Hidachi to teach him mathematics. He was given a problem as a test, which he solved so readily that the priest set about to teach Inō everything he knew in about six months. The priest then told him that if he wished to learn further secrets of mathematics, he should study under a certain scholar; to whom he sent the boy with a note of introduction. It is also said that he had already studied Chinese classics and some medical science under a certain physician at Tsuchiura in Hidachi before coming to the Inō family.

In brief, little is known of Inō's earlier life outside such fragments of tradition. In spite of his worthy birth deserving the respect of his fellow-villagers, his circumstances tended to make him looked down upon as a poor child, Sanjiro. This induced him to determine to attain distinction by hard work and self-reliance from his earliest days. That he had cherished a high ambition is evident in a passage of a letter he wrote in later years to one of his daughters. Now, the mastery of mathematics or medicine was the shortest cut for commoners to attain distinction during the Shogunate days. It is doubtless that Inō's choosing these subjects should be accounted for not only by his natural inclination, but also by the consideration just given. He lost no chance of studying in his youth, not simply mathematics and medicine, but anything whatever that would likely improve his future circumstances. It may be presumed that his study, though not very profound, extended over a wide range.

While Inō was thus helping himself, doing everything in his power to improve his circumstances, Heaven suddenly presented him with the good fortune to be the adopted son-in-law and heir to the distinguished Inō family of Sawara in the province of Shimōsa. Before entering upon the circumstances, let us describe the pedigree of the Inō family. This worthy family has a long history, its genealogy being traceable as far back as more than ten hundred years ago. The ancestors had come in 807 from the province of Yamato to settle at Inō in the Ōsuga estate, Katori County, Shimōsa, and therefore assuming the family name of Inō, had possessed the Ōsuga estate generation after generation. Towards the end of the 12th century the family lost a great part of the territory; again about the middle of the 16th century, the Inōs supporting the Kokubu family fought and lost in the battle against the Satomis; and finally settling at Sawara, they gave up their arms, and about 1573 began to live among the common inhabitants. They opened Sawara Shinjuku, and, building a new market, planned for the prosperity of the locality. The ancestor that came to Sawara first was called Kagehisa Inō, and commanded the respect of the inhabitants on account of the great services he rendered them. Since the establishment of the Tokugawa Shogunate, the Inōs, generation after generation, ranking above the common people as the chief family at Sawara, took a more prominent part in the village affairs. They were also engaged for a long time in reclaiming rice-fields along the shores of the Tone River, and made it a rule of the family to work for the public weal and to relieve the destitute. This won the reverence of the villagers and the high consideration of the authorities for the Inō family. When Nagayoshi Inō died in 1742, the only child to succeed him was a year-old daughter. It was arranged that until the girl grew up, she and her mother should be taken to the Hirayama family in which the mother had been born and that the household matters of the Inō family should be managed by their relatives. When the daughter grew up, she returned home and in 1755 married a man named Kageshige Inō. Two years later, and before he succeeded to the property, however, the man died, leaving his wife with child. On account of repeated bad harvests about 1750, the family had relieved the needy according to the family practice; and, moreover, the ill-management of the business by the relatives had decreased the revenue. Thus the family had gradually declined financially. On the death of Kageshige Inō, it was evident that before everything a man capable of reconstructing the business was needed. Suetada Hirayama, the elder brother of the widow's mother, a second cousin to Tadataka Inō, then recommended Inō to be the successor, because he recognized the latter's rare abilities in his skillful handling of a company of working-men while engaged in a certain enterprise. It was true that the social rank of the family in which Inō was born was by no means low, but that Inō himself was but a youth who had spent years in wandering from place to place might have been a ground for inward objection on the part of the widow and her mother for accepting Inō as the new master of the chief family at Sawara. On reaching an agreement, Hirayama and others planned to raise Inō's position and improve his personal connection. Thereupon, it was arranged that, instead of receiving him from his family directly into the Inō family, he should first be adopted as a son of Hirayama and then be sent thence to the Inō family as the bridegroom elect. Inō also was allowed to call himself a pupil of Daigaku-no-kami Hayashi, Superintendent of the Shōhei-kō (Academy) at Yedo, and had his name changed to Tadataka, a name that Daigaku-no-kami had chosen for him.

On January 21st, 1763, Inō succeeded to the family line as the chosen husband of the widow, and, giving up the old name of Sanjirō assumed the new one—Tadataka Saburōemon Inō. Saburōemon was the hereditary name every master of the family assumed, and Tadataka his individual name recently given him by the Superintendent of the Academy. Deeming it his first duty to retrieve the declining fortunes of the family, Inō, on the one hand, did away with ostentation, practised economy, and cut down unnecessary expenses; and on the other hand, improved the brewery business which he had inherited, besides opening a rice market, and establishing a wholesale fire-wood store in Yedo. In all this he showed steady but shrewd business ability, and worked unceasingly, setting an example to other members of the family. Consequently the family conditions improved, and the business prospered day after day; so that, in spite of giving away a great deal of money and rice at the time of the bad harvest in 1766, Inō gradually recovered the family property that had long lain waste; and before he was quite thirty, he had accomplished the greater part of his ambition.

As soon as Inō became head of the family, he was recommended by the villagers as an adviser to the headman of the village; but his exclusive attention to the restoration of the family property usually kept him out of village affairs. Whenever trouble arose, however, and he was consulted, Inō showed uncommon abilities. He came up to Yedo on business quite frequently, and it is possible that he occasionally paid his compliments to the Hayashi family; but the relation between Hayashi and Inō as teacher and pupil remained merely nominal because it is certain that the latter never studied under the former. What most pleased Inō, who even then was a lover of learning, was the numerous books in the possession of his family, but it is probable that in his eagerness to restore the property he found little leisure to study these. Tradition says that he would be absorbed in his business affairs all day, and then sit up to read after everybody had gone to sleep. It should not be supposed that he never participated in village affairs, or that he paid absolutely no attention to his studies, but until he was about thirty, his efforts were chiefly used to rebuild the business and amass a fortune, and, as it were, he was gathering latent power for coming to the fore in other ways in the future.

As Inō reached his thirtieth year, his task of replenishing the family fortunes was steadily being accomplished, and he now found some leisure for the pursuit of letters. He assumed the name Shisei as his sobriquet, and took Tōga for his pen-name. There exists a certain record which leads us to imagine that from about this time he found leisure to amuse himself, together with some literary friends, in composing poems in Chinese style, short poems, and comic verses. Seiyen Kuboki, for instance, who in later years greatly assisted Inō by writing small characters in the maps which Inō plotted after survey, was the village headman of Tsunomiya, the village next to Sawara, who being well versed in Chinese classics had struck up an acquaintance with Inō about this time, as a fellow-student of Chinese classics. For about three weeks in the summer of 1778, Inō took a sight-seeing trip to Matsushima with his wife and two attendants; a brief journal of the journey still exists, from which it may be supposed that his tastes lay chiefly in the sphere of poetry and literature, and that he had not yet developed his love for astronomy, geography, etc. The fact that he could leisurely enjoy the beautiful scenery of the north in company with his wife, despite the difficulty of communication and the extreme difficulty of a woman's travelling in these days, suggests, on the one hand, the mutual affection and tenderness of husband and wife, while, on the other hand, it shows that Inō's task of replenishing the family fortunes was nearing completion.

The village of Sawara had been ruled directly by the Shogunate until the year in which Inō travelled to Matsushima, when it was transferred to Hyūga-no-kami Tsuda, an immediate vassal of the Shogun, and became his estate. Thereafter Inō was more and more trusted by the landlord and he gave much attention to the welfare of his village; he always intervened to promote good understanding between the landlord and the villagers; in case of a calamity or an accident, he was the first to devise remedial measures, and did his best in relieving the poor; so that he became more and more popular, and was commended by the landlord.

In the summer of 1781 Sawara was damaged by a flood and the rice crop was bad, so the landlord made Inō headman of the village and ordered him to take remedial measures. Though Inō had received the title of adviser to the headman from the time that he succeeded to the Inō family, it was only, as had been customary, through the recommendation of the people. Now he began to discharge his duties at the request of the landlord. When as the result of a great eruption of Mt. Asama in 1783, falling ashes did damage as far away as Sawara, and another flood came, to make the matter worse, and a famine contained till the following year, Inō gave considerable money from his own saving for the relief of the sufferers. When the poor were employed in the construction of embankments at government expense, he busied himself with buying materials so cheaply that he was able to provide an emergency fund with the accumulated surplus. Thus, as his labour in the interest of the people was so prominent, the landlord commended him by giving him treatment similar to that given to a samurai. As already mentioned, the prestige of the Inō family was very high, and it was also respected by the authorities at the beginning of the Tokugawa Shogunate. At the time of Inō's succession, the family still commanded high respect from the villagers in recognition of its virtue through the years. However, in view of the apparent loss of its property, its standing and honour in the eyes of the authorities could not have been quite the same. Inō, who it had been hoped might relieve the Inō family, had completely fulfilled his duty and recovered not only the family property, but also the original position of the family in the eyes of the authorities. His own satisfaction may well be imagined. However, Heaven was not pleased to pour blessing upon blessing on the Inō family. Before long he had the great misfortune of having his wife die of illness.

In the fall of 1784, the landlord excused Inō from headmanship, and gave him the real power to oversee all the village officials in their management of village affairs; and within a few years an opportunity presented itself for him to display his rare ability and obtain a great reputation. There was a great flood in the summer of 1786 which resulted in bad crops and a terrible famine. The village of Sawara is upon poor, damp ground along the bank of the Tone River, and naturally it had been often visited by disastrous floods. However, the demaged districts had never been so extensive but that remedial measures had been readily devised. But as the disaster this year extended all over the Kwantō region, and as it came after the continuous bad crops since 1783, even the powerful clans well provided against famine failed to offer sufficient relief to the sufferers. The estates of the immediate vassals of the Shogun were generally given no help at all. So the people died of hunger in the fields, and a large number of them wandered from place to place. To make the matter worse, there raged an epidemic from which many died in a most miserable state. The price of rice increased by leaps and bounds until in 1787 there broke out in Yedo and other places a riot known as "the pulling-down of Temmei," Temmei being the name of this era. At this time Inō expended much of his own money in relieving over a thousand families in his own village, and also helped many others in the neighbouring villages. Moreover, he had anticipated from his own past experience and knowledge of medicine that an epidemic would surely prevail under these circumstances and had distributed free medicine among the villagers. Lest the sufferers should attempt to eat unripe corn, he took special preventive measures in other ways; and gave much attention to the health of the villagers. He bought, in advance, great quantities of rice in the Kwansai region where crops had been comparatively good, and this he had sent to be sold cheap to the rice-dealers in his village and neighbourhood, to meet the requirements of the people. He loaned funds to pawnshops so that the sufferers might get money by pawning their goods, and persuaded the wealthy people in the village to donate towards relief measures, and in many other ways, Inō did what he could in helping the sufferers. Because of this, there was not a single case of starvation nor a sign of uprising in his village. He had faithfully discharged his duties. It is also said that in spite of the amount of money he spent in relief work from his own pocket, he had part of the rice he had bought in the Kwansai region shipped to Yedo where the price of rice had risen so high that through its sale he gained back what he had lost in giving to the sufferers: he had accomplished this great work of charity without decreasing his own wealth. His lord rewarded these services in a special way. Inō frequently loaned money to the landlord to relieve him from difficulties; and in 1792 the latter rewarded him with an allowance of the rations for three men.

Before this Inō's eldest daughter had grown up and married, and was keeping house in Yedo where her husband was engaged in the rice trade. Kagetake Inō, his eldest son was already twenty years old and able to discharge the family duties by himself; and besides, as Inō had fulfilled his purpose of restoring the family property, he had an intention of resigning his official position and of retiring from active life, turning over his family business to another. Although he had frequently intimated this intention to the landlord, the latter had not given his consent because he had thought the villagers not quite recovered from the famine and Inō's abilities would still be needed.

However, since Inō left practically all his family business with his eldest son and had no great concern from this time on, he had leisure to indulge in learning according to his own taste. And what most attracted him now was, not the versification he had been so keen about ten years before, nor any other line of literature, but an eager study of astronomy. Such a change in his taste did not take place suddenly from any particular motive. As he had always liked mathematics from childhood, it is certain that he had long been more or less interested in some calendaric computations, then considered as a sort of collateral of mathematics. Especially, since the solar eclipse in the fall of 1763 was miscomputed in the state calendar and the incompetence of the calendar officials was more and more ridiculed among the people, and a great number of astronomical books were written, all this must have roused general curiosity to study astronomy, and it is possible that Inō's interest in this direction was also thus stimulated. But this study of astronomy was not to be mastered during the spare moments of his business, like the composition of brief poems. Therefore, while he was busy, he amused himself with composing poems; now that he found ample time for study, he was engaged in the study of astronomy, his cherished desire from childhood. A study of the fragments of his letters and other documents of this time will indicate that he first bought some astronomical books and studied them for himself. But most astronomical books then accessible to the public were based upon the Shou-shi Calendar and, lacking in details and accuracy, were full of doubtful points; the more books he bought and studied in order to clear these doubts, the more interesting to him the subject became and the more questions arose in his mind. This led him, it seems, to plan a fundamental study of the subject and to clear up every doubt he had entertained.

During the period that Inō was thus absorbed in teaching himself the science of astronomy, one or two of his personal affairs seem to be worthy of note. One is that in the summer of 1790 he married as his second wife the eldest daughter of Ryūchō Kuwabara, a physician to the Sendai clan, then stationed in Yedo. Little is known of Kuwabara's career, or of the circumstances under which Inō came to marry his daughter. Since Kuwabara professionally visited the residences of high officials, he had a wide circle of acquaintances. Later on, when Inō planned a survey of Yezo, Kuwabara was instrumental in gaining inside information regarding the intention of the authorities and became an able adviser to Inō, along with Yoshitoki Takahashi, and contributed much towards the realization of Inō's enterprise.

Another matter of note was an extensive journey of more than three months through the Kwansai region in the spring of 1793. According to his own record, his chief object in this journey was to visit the Great National Shrine in Ise. Incidentally he visited all the noted places and historical sites along the Tōkai route and, after arriving at the Great National Shrine, went through the Kinki section. Then on his way back, he took the Kiso route because of its beautiful scenery. The journey enlarged his experience and increased his information considerably. His journal also reveals that his taste for science was keen at that time and his observation or point of view during the trip was characteristic of a scientific spirit. When he came to a beautiful place, commanding an extensive prospect on every side, he took out his pocket magnetic compass and roughly determined the azimuth of each spot he was to see and studied the natural features of the locality; and when the sky was clear enough at night he determined the latitude by roughly measuring the altitude of Polaris. The results of these efforts are all recorded in his diary, and they prove that Inō had a special taste for geography, land-surveying, etc. even before he came to Yedo to study the new sciences.

In 1794 Inō again implored the lord to let him resign his post and retire to private life and obtained the latter's consent at the beginning of 1795. He then formally handed over his business to his heir, Kagetaka Inō, and retired from active life. He gave even his hereditary name, Saburōemon, to his son and changed his own name to Kageyu. After this, therefore, he usually went by the name of Kageyu and only when a formal signature was necessary, he used his true name Tadataka Inō. Finding himself free both in name and in fact, he planned to find a quiet home in Yedo and penetrate into the secrets of astronomy under a competent teacher. Unfortunately his second wife, of the Kuwabara family, who had been undergoing treatment at her own home in Yedo, died in the spring of 1795, leaving him a widower a second time. Therefore, his son protested against his sojourning alone in Yedo and did his best to make him change his mind and to live at Sawara to enjoy the remainder of his life there. But he positively declined, and at the age of fifty went up to Yedo as student in pursuit of knowledge and took up his residence on Kuroye-chō in Fukagawa. This place became the base station of his various astronomical observations and also a most memorable spot as the headquarters of his survey of Japan. Although on account of city improvement in recent years the place has undergone a great change in streets and canals and is now almost beyond recognition, its position may be located pretty accurately from Inō's record—namely, the north-western part of No. 29 of Kuroye-chō, as it was before the great earthquake of 1923.

How great a degree of scholarship he had attained by this time cannot be clearly known now, but it may be roughly inferred from the fragments of measuring-instruments which are alleged to have been used by him before his coming up to Yedo, by the catalogue of his reference books, and by other fragmentary documents.

Mathematics formed one of his favourite studies from childhood; and tradition says that he had studied it rather deeply before he married into the Inō family. Now, the Kadzusa-Shimōsa region had so many masters of mathematics that it was possible for a youth, by imposing upon the hospitality of his relatives and friends, to study under one of them. We should credit this tradition. There is no doubt that, when he reached fifty, he had mastered most of the advanced courses in Japanese mathematics. In the catalogue of the books alleged to have been in his possession before his coming up to Yedo, we find not only the "Shūki-sampō," the "Kaishō-tempei-sampō," and many other books on comparatively advanced mathematics, but also the "Taisei-sankyō," the book containing some of the secrets of the Seki school. The presence of this last proves his acquaintance with the more advanced mathematics. Whether he received the above-mentioned work after taking a complete course at a Seki school it is impossible to determine on account of the absence of data.

His knowledge of astronomy, it seems, was limited to a careful perusal and comparison of the theories given in the astronomical books then accessible to the people as already stated. And in order to discuss the different theories carefully, it was necessary for him to test them not merely by past astronomical phenomena, but also by his own obversations. Yet Inō made no such accurate observations, it seems, while he was at Sawara. The astronomical instruments which we think Inō employed before his coming up to Yedo, as we infer from their fragments and other articles still preserved by his descendants, were only a gnomon, a magnetic compass, and a simple vertical circle graduated to a half-degree—all unfit for making an accurate observation.

All his knowledge of land-surveying up to that time was chiefly what had been published in the "Ryōchi-shinan," the "Kiku-buntōshū," and the "Chōken-bengi" which are included in the catalogue of the books in Inō's library. His connection with the village affairs of Sawara and the necessity of defining the boundaries of the farms of the village induced him to make several maps of the village. These, together with the plans of his dwelling drawn in 1792 and 1794, are still preserved. There also remains a wooden theodolite which seems to have been used by him. From these relics it is evident that he was already familiar with the principles of land-surveying and was as competent as the ordinary land-surveyor or so-called draughtsman of that time. In a copied volume of the book entitled "Kikujutsu-denrai-no-maki," Inō is mentioned as a pupil of Toshiyori Matsumiya. Since this book is full of other questionable points, this statement is hardly credible. In no other document now existing is any mention made of a relation between Matsumiya and Inō as teacher and pupil.

We may incidentally note Inō's attainments in the science of medicine and in Chinese classics. All he did in the study of medicine was possibly the reading of some books of the old school and he had no ability to practise medicine generally. As to his study of Chinese classics, it is possible that he educated himself with the help of various books and also as a result of his study in company with friends, he no doubt had already attained a fair grade of scholarship when he came up to Yedo. However, he had not studied the subject properly under a competent master, and he never planned to get on in life with his knowledge in this line. That his mastery of Chinese, frequently overestimated by some people, was quite incomplete may be inferred from the quality of his Chinese composition in the "Bukkoku-rekishōhen-sekibō" (In Confutation of "The Indian Astronomy") which he wrote in his last years.

2 His Contact with the New Sciences

The year 1795, in which Inō came up to Yedo for the purpose of studying astronomy under a competent teacher, was also the very year in which Yoshitoki Takahashi and Shigetomi Hazama came to Yedo by the order of the Shogunate. It was as though Heaven, being pleased with Inō's desire, had provided him with these teachers and friends. On hearing of Takahashi's reputed mastery of the new astronomical theories of the West, Inō visited him; and as he listened to the master explaining part of a new theory, Inō found that his argument was so logical and his calculation so accurate as to be incomparably beyond the old theories. Inō at once made application and became a pupil of Takahashi. He had then already reached the age of fifty while Takahashi was in the prime of manhood—at the age of thirty-one.

Takahashi, in teaching his pupils, made it a rule to instruct them by a gradual process—that is, each pupil had first to master, before coming to the new theories of western countries, the Shou-shi Calendar designed in the Orient and then only was he allowed to advance to western astronomy expounded in the first and second volumes of the Li-hsiang-k'ao-ch'êng and not yet modified by the theory of the elliptic orbits; and after he thus accustomed himself to astronomical computations, he was then permitted to study the new theory of the elliptic orbits by the Sequel of the Li-hsiang-k'ao-ch'êng. Since Inō had mastered by his own efforts the Shou-shi Calendar before coming to Takahashi, he evidently commenced a study of the first and second volumes of the Li-hsiang-k'ao-ch'êng immediately following his matriculation. Moreover, Takahashi, noticing Inō's ardour for study at a comparatively advanced age, did not regard him as an ordinary pupil, but gave him specially favourable treatment; besides instructing him on the regular lecture-days, he made Inō submit a written report of the results of practical computations according to his attaiments and answered his questions—which proved of great service in Inō's rapid advancement in scientific studies.

In the autumn of 1796, or the year following Inō's entrance into Takahashi's school, his master was called to Kyōto for the important task of compiling a new calendar. During his absence Hazama answered Inō's questions for him and as to the results of the astronomical observations Inō had taken, he reported them to Takahashi and Asada in Ōsaka. Much of the correspondence between Takahashi in Kyōto and Hazama in Yedo is still extant. According to some of the letters, towards the end of 1797 Inō was studying the new theory of the elliptic orbits by the Sequel of the Li-hsiang-k'ao-ch'êng and had already progressed so far as to understand its astronomical principles and methods of computation. As he gave full play to his unrivalled energy in studying and was altogether undaunted by the many difficulties he came across, his accomplishment towards 1800 already had won him a reputation as an astronomer of the first rank among the followers of Takahashi. It must be admitted, however, that Inō shared in the benefits which such pioneer scholars as Asada, Takahashi, and Hazama had brought forth after many years' toil, as he was able to acquaint himself thoroughly with the then newest theory without experiencing great difficulties, and consequently he does not compare fundamentally with those pioneers who were at such great labor to develop the new sciences. The great man, therefore, who, utilizing this new knowledge, was able later on to perform the stupendous task of surveying all Japan, was, in the eyes of these pioneers, only a new graduate who had completed a general course of western astronomy with unusual ardour. This is quite apparent in the correspondence that passed between them.

When Takahashi obtained the Astronomia by Lalande in 1803 another great stimulus was given to the native astronomers. While Takahashi himself was absorbed in studying this and while his two sons laboured at translating it into Japanese, Inō, as will be described in the following pages, was engaged in the enterprise of surveying Japan, and being ever on the move, could not himself share in the study of this new book on astronomy. However, the new knowledge which had been gained through the efforts of his leaders and friends was freely given him at his request and thus Inō was able to improve his knowledge of astronomy until the end of his life. According to the documents extant, it would seem that Inō, in his devotion to surveying after 1800, added to his stock of knowledge only fragmentary bits of new astronomy and never succeeded in mastering the theories expounded in Lalande's Astronomia or applying them freely to practical use. Again, the meagreness of his knowledge of the Dutch language is evident. Even in his last years, he knew only the Arabic figures, the alphabet, and some terms commonly used in astronomy, but was by no means able to read a Dutch book.

While thus pursuing the study of astronomical theories, Inō spent a large sum of money in purchasing or constructing new measuring-instruments and practised various methods of observation at his house at Kuroye-chō. Since the fortunate Inō had made the acquaintance of Hazama who had a profound knowledge of measuring-instruments of the new style, he purchased a pendulum-clock through Hazama from the mechanician in Kyōto whom the latter had trained, and also had an accurate quadrant, gnomon, etc. constructed by a mechanician in Yedo under Hazama's supervision. With the help of these instruments, Inō measured the altitude and the time of the meridian passage of the sun, the moon, and fixed stars, and lost no opportunity of making an observation of solar and lunar eclipses and other special phenomena. As he worked hard day and night, his skill greatly developed. And his eagerness in making observations may be inferred from the fact that he had an opportunity to observe the meridian passage of Venus in broad day-light. This was in 1797 and Inō was the first to observe the meridian passage of Venus in Japan. After this, Inō procured more measuring-instruments of a superior quality with which he worked hard at his observation till towards the end of the century he grew more proficient in technical skill than any other of the Takahashi school. He was now an expert observer in the opinion of others as well as in his own.

Inō's great eagerness in making astronomical computations and observations during his student days under Takahashi is given in the Memoir of Inō, which Kagesuke Shibukawa wrote from his own knowledge. "Old Inō, on hearing the reputation of my father Tōkō (Yoshitoki Takahashi), came to be a pupil under him. As he was instructed in the western method, his doubts of many years were cleared up and after studying it sleeplessly, he computed a calendar and solar and lunar eclipses for two or three years single-handed only using the method and tables given in the Sequel of the Li-hsiang-k'ao-ch'êng. My father Tōkō was greatly impressed by his patience and jokingly called him Suihosensei (Mr. Computation). Since he was equipped with new astronomical instruments, he could not bear leaving the house. If he went out in the forenoon, he was back by noon to make an observation of the sun at noon; if out in the afternoon, he came home by dusk and observed the stars. He never talked at leisure unless the sky was overcast. When he talked with Tōkō on astronomy, taking no note of time, if he suddenly became aware of the oncoming of evening, he would flee in a panic for home. It often happened that he would leave behind his small sword, pocket-book, and so on. Old Inō's eagerness for making observation accounts for this. . . ." This quotation will sufficiently illustrate the point.

3 His Motive for Undertaking a Survey of Japan

The original purpose for which Inō came to Yedo and studied astronomy was to clear up the doubts he had felt while studying for himself at Sawara and to learn the true principles of astronomy. By the year 1800 he had completely mastered the Sequel of the Li-hsiang-k'ao-ch'êng, known to be the very best work on astronomy, had cleared away all the doubts of former days and become an expert in the technical work of astronomical observation. His original purpose had been achieved. However, his character was not such as to be satisfied with this and with enjoying the rest of his life leisurely discoursing on heavenly bodies. He had to be continually active in some direction or other. He now surprisingly planned a survey of Yezo at his own expense and this desire was conveyed by Takahashi, his master, to the high officials of the Shogunate. By what motive did Inō choose a survey of Yezo as his field of action?

As stated in the preceding chapter, Takahashi experienced much difficulty in his astronomical computations because he did not know the longitudes and latitudes of the more important points of this country, nor the accurate magnitude of the earth. Therefore, it cannot be denied that it was partly by his persuasion that Inō made up his mind to determine accurately the meridian length of the earth and that was of course his chief reason for beginning a survey of Japan. It would not be the whole truth of the matter, however, to attribute Inō's motive in deciding to carry out the survey even at his own expense, to Takahashi's persuasion. The importance of the determination of the magnitude of the earth in astronomical computations was not to be perceived by Takahashi alone. All the astronomers at that time desired to obtain its exact figure and Inō had long been aware of the necessity of an accurate survey. The latitude difference between Kuroye-chō in Fukagawa where Inō lived and Kuramae in Asakusa where the Astronomical Bureau (the Asakusa Observatory) in which Takahashi lived was situated was about 1′·5. This value was pretty accurately determined through the observation of the meridian altitudes of a number of fixed stars made both at the observatory and at Inō's residence. Inō wished to compute the meridian length of the earth by surveying the meridional distance between these two points, but as a mere astronomer would hardly be permitted to survey the streets of Yedo openly, the seat of the Shogun, he stealthily carried out a rough calculation, measuring the distance by pacing and the azimuth with a small magnetic compass. The sketch-map he drew on a piece of paper has not been lost and is still preserved at Inō's house. He took the result of this rough survey to Takahashi and expressed his desire to obtain a still more accurate result by carefully surveying the distance between the two points with the permission of the authorities. Then his teacher informed him of the difficulty of accurately determining the magnitude of the earth by a survey of such short distance as within the city of Yedo. He also advised him to wait for an opportunity to carry out a survey on a grand scale and to obtain results of permanent value. So it was Takahashi who magnified the scale of Inō's plan to determine the earth's magnitude.

Besides this plan, Inō had an ambition for map-construction. That Inō had long wished to survey different provinces and plot their maps is evident in a letter he wrote to the authorities of the Shogunate at the time he started for Yezo. "Now that I have studied day and night for six years as a pupil of Mr. Sakuzaemon Takahashi I am able to make measurement without error. For the purpose I have collected various instruments and expended more money than would befit my position. Though it is but the hobby of a retired old man, it often makes me feel sinful. So I have wished to make maps that might serve as reference to the coming generation. To make a survey with measuring-poles and measuring-ropes and carry about large instruments within the territories and estates of Daimios and immediate vassals of the Shogun, would certainly provoke a challenge from the government officials and it would be deemed unpardonable in a man of my position. Though I was born in Kadzusa, I married an heiress in Shimōsa and travelled often to Hidachi on business as a rice-dealer. Thus I have made a number of acquaintances in such provinces as Hidachi, Kadzusa, Shimōsa, and Awa. I was thinking that if I were to stay with them, I could make maps of small sections. Fortunately, now I am leaving for Yezo, recommended by the Shogunate. . . . ." However, Inō's motive in planning a map-making of the provinces should not be attributed, as expressed in this letter, solely to his desire to utilize the measuring-instruments which he had had constructed for astronomical purpose. The map of Japan which Genshu Nagakubo had made in 1778, though only a small one and not based on actual survey, had met the approbation of the public and Nagakubo had been created a samurai because of this service.

This fact, of course, had stimulated Inō's pent-up ambition and formed one reason for his resolution to carry out the unprecedented enterprise of this wide survey. He also naturally desired to win a reputation and to harvest the results of his labours of many years.

The situation of the world in the latter half of the 18th century had undergone a remarkable change from that in the first half of the 17th century when Iyemitsu Tokugawa proclaimed the isolation law and strictly prohibited the people from communicating with any western country except Holland. Such foreign countries as Spain and Portugal, harbouring sinister designs upon Japan at the time of the declaration of the isolation law, had gradually become less powerful; and Great Britain was so busily engaged in colonizing North America and trading with India that she had no desire to communicate with Japan. Therefore, the strict law was enforced without difficulty, and for the succeeding two centuries the Japanese enjoyed the blessings of peace. In the mean time Great Britain gradually prospered, and in spite of the fact that the United States of America, then her colony, shook off the yoke of the mother country and became independent in 1783, she gained tremendous power in India about this time and designed to extend her power into the south-eastern part of Asia. Russia, too, since the accession of Peter the Great towards the end of the 17th century, had suddenly changed her policy and expanded her territories by invading her neighbouring countries, encroaching upon Rashau and Simushir, two of the Kurile Islands in 1765. In the following year some Russians settled in the island of Urupp. After this Russian ships more and more frequented the northern seas and were eagerly watching for a chance to invade our northern frontier. At last, in 1778 a Russian ship came to Yezo and applied for the opening of commercial intercourse and came back again the following year demanding an answer. Hereupon, patriots awoke to the necessity of defending the coasts; for instance, Shihei Hayashi published the "Sangoku-tsūran" (A General Survey of the Three Countries) in which he discussed the geography and conditions of Japan and the neighbouring countries, and again the "Ķaikoku-heidan" (Warfare for an Insular Country) in which he advocated a policy of national defence and awoke the public to its danger. The Shogunate, displeased with these books, imprisoned the author. But, in the face of the progress of the times, it could no more enjoy the slumber it had had during the two preceding centuries. The government now sent Tokunai Mogami and Jūzō Kondō to explore Yezo and plan a policy of colonization, and then placing the eastern part of Yezo under its own immediate rule, ordered its officials to investigate the question of garrisoning the territory. Thus the situation in the north appeared gloomy and threatening and was ominous of an imminent stirring throughout the whole country. As for maps, very necessary for planning the defence of coasts, there existed none made on the basis of an actual survey, even for the Main Island; there was not a single one to be relied upon for Yezo. Nisuke Hotta, a samurai of the Tsuwano clan of Iwami and assistant in the Astronomical Bureau of the Shogunate, was sent in 1799, by boat, to Yezo for the purpose of making a coast-line map from Yedo to Yezo, but he returned to Yedo without accomplishing much, except the making of a few rough maps. It is only natural that Inō at this juncture should have petitioned for permission to survey Yezo at his own expense, so that he might, on the one hand, throw light upon the topography for the need of the government and, on the other hand, determine the magnitude of the earth for scientific use, thereby also making a name for himself. Thus, Inō's survey enterprise was by no means incited by a single motive as might be supposed, but by a composite one, comprising his taste and ambition, Takahashi's persuasion and intercession, astronomical and geodetic necessities and the needs of the times.

Still another word is necessary as to why a survey-map should have been made by Inō, but not by any one before him. In spite of the fact that the methods and instruments for surveying a tract of land for which the surface of the earth might be considered as a plane were known for a long time, and that even in making a survey of a more extensive area, a method to avoid the accumulation of errors by means of observing the astronomical longitude and latitude was known, yet to the disappointment of the public, no survey-map had been made in Japan until about 1800. The following reasons might be given. (1) That the so-called land-surveyors were mostly mere theorists, seldom gifted with real ability to apply their knowledge to practical affairs freely as the occasion demanded. (2) That, among ordinary land-surveyors, almost none was capable of accurately measuring astronomical longitude or latitude. (3) That the measuring-instruments were so coarsely constructed that their errors were gross; especially the portable astronomical instruments were like toys and not suited for an accurate survey of an extensive area. (4) That it was impossible to make a free survey in the estates of feudal lords unless by order of the Shogun or under special circumstances.

Now, Inō had not only studied western astronomy and mastered the art of astronomical observation and land-surveying, but also possessed various measuring-instruments precise enough to surmount most of the difficulties his predecessors had encountered. Inō was also the person best qualified for the task of surveying the country. However, if he had not secured the permission freely to survey the estates of feudal lords, his technical skill could not have been developed in this direction. And it was most fortunate for Inō—nay, for our country that Inō came forward at the very moment a survey-map was most urgently needed by the Shogunate and, with its support, he embarked upon this task of survey.

4 The Progress of the Survey

(*A*) *His Negotiations with the Shogunate*

It is not clear as to when Yoshitoki Takahashi first informed the Shogunate authorities of Inō's desire to survey Yezo. According to Inō's diary, it seems that, at the beginning of 1800, Takahashi broached the subject to the Junior Secretary and some subordinate officials and step by step entered upon a series of negotiations. As the officials in charge decided to grant Inō's petition and intended to make Inō go by sea to Yezo, they sent men to his house to investigate the volume and weight of his surveying-instruments and additional baggage. The reason why the officials chose the sea route was to facilitate the transportation of the baggage and to obtain, if possible, a rough topographical survey of the eastern coast of the country, taken from the boat while sailing. It is clear that if he obeyed the will of the Shogunate officials and took ship both ways, Inō would naturally have to give up his plan to survey, on his way to Yezo, the distances and azimuths along the Ōshū Highway and the latitudes of the more important points. He had hoped to make a survey-map of the route and thereby determine the length of one degree of the meridian which he had wanted so badly in his astronomical computations. Therefore, Takahashi at once reported to the Junior Secretary, submitting a letter in which he gave the reason why a trip by land was necessary, and also reported roughly the number of carriers needed for conveying the surveying-instruments by land.

In spite of urgency, no answer was received for nearly a

month. As Inō intended to leave Yedo early in the spring and come back before the cold season set in, he was much irritated. Ryūchō Kuwabara, his father-in-law, on secret investigation, learned that the attitude of the officials of the Shogunate was one of indecision and procrastination because of the cost of so many carriers, in case the land route be adopted. Thereupon, Takahashi intimated to the Junior Secretary that Inō, being confident that the permission for the survey was forth-coming, had expended a considerable sum of his own money in constructing a number of new measuring-instruments, and petitioned for the speedy granting of the official permission. Concerning this, the official in charge of Yezo affairs sent word to Takahashi to the effect that owing to the frequent demand for men and horses along the route in the Main Island as well as in Yezo, since many had to travel on official business, it was impossible to accomodate Inō with as many men and horses as he desired, and that he should either have his baggage shipped to Yezo or, if he should take the land route, he must decrease the number of men and horses as much as possible.

Inō then personally negotiated with the official in charge of Yezo affairs on the matter of decreasing the number of men, and after he cut down the number of measuring-instruments to be taken to Yezo to the minimum, an agreement was at length reached:—that Inō be authorized to hire not more than 6 men and 2 horses, at his own expense, though in the name of the Government. On Inō's proposal to defray all the expenses of the survey, the Government authorities considered it proper to grant him some allowance, as the result of the survey was to be taken over by the Government. They decided to give him 7·5 *momme* of silver per day while travelling on official business—which was the amount received by a common practising physician when temporarily employed by the Shogunate.

So it was that in June, 1800, Inō was given a written appointment at the official residence of the official in charge of Yezo affairs. It read: "In accordance with your persistent desire, you are ordered to proceed to Yezo for a surveying test. You are expected to work carefully. While you are on service, 7·5 *momme* of silver per day shall be allowed you." It was somewhat of a privilege that Inō should have been given an allowance, though relatively small, and that he should have been treated, not as a retired farmer, but as a lordless samurai. However, this cannot be regarded as a sufficient facility for the Government to accord to his enterprise. The fact that the Government hesitated several months before giving consent to Inō's proposal when an accurate map was pressingly needed, and that it gave him such scanty support, betrays a lack of wisdom on the part of the Government officials to adopt a timely measure, as well as their lack of confidence in Inō's abilities and their doubts as to the results of his survey. This was perhaps because no surveyors sent to Yezo previous to this had fulfilled their mission. The phrase "for a surveying test" in the written notice of appointment may suggest these circumstances.

(*B*) *The Survey of Yezo*

Inō had made all preparations to carry out the survey and was waiting for the issue of the order. On receipt of the permission, therefore, putting on his travelling-clothes at once, he took his first step from his residence in Kuroye-chō, Fukagawa, entering upon the enterprise of surveying the whole country, on June 11th, 1800. He was already at the age of fifty-five. Those who followed him were Shūzō Inō, his second son, Hayata Kadokura, Sōhei Hirayama, and two servants. Although the minute record of the measuring-instruments he took with him has not been preserved, it may be surmised from the manner of his survey and from his survey-journal that they were a quadrant 1·2 m. in radius, a magnetic theodolite 0·75 m. in diameter, a pendulum-clock, a small magnetic theodolite mounted on the top of a stick, common magnetic compasses, measuring-rods, measuring-ropes, telescopes, etc.

After leaving his residence, Inō took the Ōshū Highway, on which he measured the distances by constant pacing; and he roughly measured with the stick-theodolite and other instruments the direction of the road and the azimuths of distant mountains, making 30–50 km. daily, the days then being long; and when he arrived at an inn, he set up the quadrant and determined the latitude of the point by observing the meridian altitudes of the fixed stars. In this manner he and his party reached Mimmaya in Ōshū in about 3 weeks, having passed through Utsunomiya, Shirakawa, Sendai, Morioka, Noheji, Aomori, etc., and then with a favourable wind they sailed for Yezo and arrived at Hakodate. In those days the island of Yezo was in an almost primitive state and inhabited only by Ainos engaged in fishing, and officials sent by the Shogunate to various scattered stations along the coast for the investigation of defence and colonization; the roads were exceedingly rough and dangerous; for lodging they had only meeting-houses or guard-houses, built 20 or 30 km. apart. The lack of men to carry the measuring-instruments compelled Inō to leave the large magnetic theodolite at Hakodate, and the only bulky instrument he was able to take with him was the quadrant. At first Inō had planned to survey the coast-line of Yezo with the measuring-ropes, but fearing that this method would take so much time as to hamper the anticipated progress of the survey, he now decided to measure the distances in Yezo also by the method of pacing. He surveyed the south-eastern coast of Yezo from Hakodate, making 20 or 30 km. daily, and making astronomical observation at each meeting-house or guard-house where he put up, and whenever the weather was bad, he would stay until he could make observations and determine the latitude of the station. He succeeded in surveying the route without interruption as far as Hosenji, having passed through Yamakoshinai, Muroran, Mitsuishi, Horoidzumi, Ōtsunai, Kushiro, etc. After crossing Akkeshi Bay by boat and landing at Akkeshi, he left the coast by way of Annebetsu for Nishibetsu, not far from Nemuro, the eastern extremity of the island of Yezo. The difficulty of the road hitherto compelled him to rely upon a river-boat so frequently that his survey of the distances or azimuths along the route became quite inaccurate. Therefore, the position of Annebetsu or Nishibetsu was determined only by making use of the astronomical latitude of that place and azimuths of the two high mountains, Oakan and Meakan, sigted from various stations. Though Inō attempted to survey from Nishibetsu to Nemuro, it was difficult to get men to carry his instruments, because of the coming of the salmon-fishing season. Moreover, as more than two months had already elapsed since leaving Hakodate and now the autumn chill was felt, he started back upon his return journey, after roughly determining the position of Nemuro and the capes of the island of Kunajiri by observing their azimuths from distant points near Nishibetsu.

He chose the same route on the way back, and measured the distances and azimuths again so as to confirm the measurement he had taken on his first journey. As he did not need to stay more than one night for astronomical observation except at one or two places, he reached Hakodate in about half the time that his former trip had taken. He surveyed the road westward from Hakodate as far as Matsumae, and took ship there to cross the strait to Mimmaya. Then, as on his outward trip, he roughly surveyed the Ōshū Highway and made latitude observation at each station where he spent the night. Thus he proceeded on his way home. Because of the shortness of days during this season, the distance covered each day was much less than before. On the trip down from Mimmaya he covered about 25–35 km. daily, and reached Yedo on December 7th, having spent more than a month on the way.

The Shogunate Government did despatch instructions to the village officials along the route, telling them to accommodate Inō, without delay, with the necessary men and horses that he might require, at the fixed rates, and to make him feel no inconvenience in the way of ferrying, fording, or lodging. During the journey, therefore, Inō was only exempted from the fares for ferrying or fording; he paid at the fixed rates of an ordinary traveller, not only for lodging, but also for the men and horses that were needed in carrying his surveying-instruments. Though he tried his best to save expenses, the sum he was required to expend out of his own pocket besides the 22·5 *ryō* allowed by the Government, amounted to 80 *ryō* (about ¥1000 of to-day). However, Inō was considerably pleased with the few facilities and considerations given him upon this journey; otherwise, as soon as he had left home, he would have had to give the way to any passerby of the samurai class, and to experience many other forms of humiliation.

After he came back to Yedo, he occupied himself in plotting a map of the regions he had surveyed, being assisted by the members of his party; Seiyen Kuboki, a friend of his, and a certain woman by the name of Yei. Though the woman was really his third wife whom he had married in 1797, because he was then over fifty and a man retired from active life, he dared not call her his wife in the face of the world, but treated her as a special member of the family. She had profound learning, and was good at Chinese classics and mathematics, skilful in manual labour such as graduating the circle, and also in drawing. It is possible that she rendered much help in Inō's planning to survey

Yezo, but unfortunately documents to enlighten us on this subject are lacking. Inō's survey of that year had not been planned to be highly accurate, and consequently the work of mapping was so simple that the two kinds of maps in different scales were completed within about two months. The larger consisted of 21 sheets plotted on a scale of 1:43636, and the smaller was one sheet map of a reduced scale of one-tenth of the larger. To these maps there was an appendix in which the methods of survey and plotting were explained. These were presented to the Department of Yezo Affairs, and a copy was given the Junior Secretary. They expressed their satisfaction at Inō's rare skill in land-surveying. According to the appendix, Inō had obtained for the length of one meridian degree an approximate value of *27 ri* (108 km.) as the result of his latitude measurements and his pacing from Yedo to Noheji for a distance of some 5 degrees in latitude difference; and he had plotted the maps according to this approximate value.

As the survey of Yezo was the first occasion of putting his knowledge of land-surveying to practical use, and also a test, as it were, of his technical skill, given by the Government authorities, Inō might have suffered tremendous mental strain while travelling; this in addition to the physical difficulties. It was fortunate that when he made such a trip with a group of young inexperienced people to and from a remote place difficult of communication, no one was taken ill, and that he was able to fulfill his mission. Though a few things contrary to his expectations had happened, as such always do at the commencement of a new enterprise—such as the necessity of leaving his large magnetic theodolite unused at Hakodate, the inability to survey all the route with measuring-ropes as he had originally planned, or the exceedingly inacurate survey east of Akkeshi—it was a matter of great gratification that he had as a whole fulfilled his purpose.

Even Yoshitoki Takahashi, a man who knew Inō's abilities best, had had doubts as to whether he would succeed or not when he left for Yezo. Therefore, when Inō returned with greater success than he anticipated, Takahashi was overjoyed at the thought that his instruction of long years had not been in vain after all. He at once sent a letter to Hazama in Ōsaka, in which he reported the processes of Inō's survey of Yezo. "After leaving Yedo," Takahashi wrote, "Inō measured latitude at every point where he stopped for the night, and such observation was continued at each stopping-place in Yezo also. What impresses me is his immense labour in marking the number of paces between Yedo and a place named Nishibetsu, a distance of about 420 or 430 *ri,* without leaving out any section of it. He determined the direction of the route by measuring the magnetic azimuth with a compass 3·5 *sun* in diameter. He also surveyed the azimuths of high mountains at various places. All such surveying I had told him to make, it is true; but I had hardly expected him to accomplish it so thoroughly. I cannot help admiring his achievements. . . ." Takahashi's joy may well be imagined.

In connection with Inō's survey of Yezo, there is one thing that should be recorded. This is his chance meeting and the forming of a friendship with Rinzō Mamiya, while engaged there in his survey. As Mamiya was then only a petty official, he failed to impress Inō much. This chance meeting, however, led to Mamiya's coming to Inō to practise astronomical observation and land-surveying under him, and Mamiya put his attainment to practical use in Yezo. He later on offered the materials of his own survey to Inō, and these the latter used to supplement the sections of Yezo that he had failed to survey.

(*C*) *The Survey of the Eastern Coast of the Main Island*

Inō had intended to spend two or three years more in completing a survey of the coast of the island of Yezo, and also in surveying the eastern coast of the Main Island between Ōshū and Yedo, making a perfect map of the sea-coasts from Yedo to Yezo. So as soon as the mapping of Yezo according to his tentative survey was finished he intended to realize this desire, and, after holding a conference with Takahashi and Kuwabara submitted a petition and plan to the Shogunate. According to this plan, Inō was to survey the eastern coast of the Main Island as far as Mimmaya, and then sail to Matsumae in Yezo, thence surveying the coast clock-wise to Nemuro. He was then to survey such islands as Kunajiri, Etrofu, and Urupp, and besides, determine the longitude astronomically by observing the lunar eclipse on Sept. 22nd of that year on one of these three islands. On his way back, he was, in order to revise the map, to re-survey the section he had measured quite roughly the year before. Inō was determined to take various instruments in order to secure an accurate survey this time, and demanded that enough men and horses be furnished by the Government for carrying these instruments. He also demanded that a boat be purchased at Matsumae at the Government's expense, to follow the survey-party with materials for hut-building on land and with provisions, so that wherever traffic was impossible on the north-western coast of Yezo, the boat might be helpful in carrying out the survey. Before Inō formally submitted the petition, Kuwabara informally showed it to the officials in charge and asked for their opinion. The furnishing of men and horses for carrying the instruments and the purchase of a boat by the Government seemed not likely to be granted. Takahashi and Kuwabara, therefore, advised Inō to strike out these two clauses from the proposal before it was submitted. Inō would not listen to them for the elimination of the two clauses would surely result in an inaccurate survey, and all his travel through wild unhealthy regions, and his toils and pains, would be in vain. This plight cost Takahashi tremendous trouble. Finally he advised Inō that though an accurate survey of the north-western coast of Yezo might be impossible without the special co-operation of the Government, that of the eastern coast of the Main Island would not be difficult at all, and that if Inō were willing to work at the same allowance and with the same treatment that he had received the year before, he had better apply only for the survey of the eastern coast of the Main Island. To this Inō finally consented and striking out from the original plan the survey around Yezo as well as the two items, decided on a new plan, to go from Yedo to Mimmaya along the eastern coast of the Main Island, and to survey a little around Matsumae in Yezo; then on his way back, to take the Ōshū Highway in order to revise the preceding year's survey. This he referred to Takahashi to have submitted.

An event which did Inō great honour occurred about this time. When he had resigned his official position at Sawara, Kagetaka Inō, his eldest son, had been appointed his successor, and Kagetaka had fulfilled his duty in improving the welfare of the poor. The villagers were devoted to Inō and his son for their virtue, and regretted that the good deeds of the Inōs had never been recognized by the Government authorities. About the beginning of 1800 the villagers of Sawara and some from the surrounding villages drew up a testimonial of Inō's good deeds, done during the famine of the Temmei era, and also of the commendable conduct of the Inō family generation after generation. This they submitted to the Supreme Court for investigation and recognition. Accordingly, the officials often summoned the people concerned and inquired into the facts, which they finally decided were deserving of recognition, as the villagers had requested. In March 1801, Inō and his son were summoned to the office, and were highly commended for their good deeds, each being given ten pieces of silver. In addition, they were granted special permission to be known by their surnames and to wear swords. It is true that Inō had already been given treatment equal to that of a samurai from the lord of Sawara; and during his journey to and from Yezo the year before, he had been given the treatment of a lordless samurai, but that was a mere recognition of a local lord, or a mere temporary expedient on the part of the Government. Now Inō was formally given the treatment of a regular samurai by the Shogunate in recognition of his good conduct in earlier years. And such a privilege was so rare that it goes far in attesting the nobleness of Inō's character. No doubt it also proved quite advantageous to him in his negotiations in connection with the plan of survey.

At the same time Takahashi submitted Inō's petition to the Department of Yezo Affairs. Now, this plan had to do with Yezo, it is true, but the land which Inō proposed to survey was part of the Main Island, and some negotiations with the Commissioner of Communication were necessary. After some parleying, the districts to be surveyed that year were changed to the coast of Idzu and the other provinces along the eastern coast of the Main Island as far as the province of Mutsu, and the official sanction was issued in April. The formal assistance Inō received was about similar to that he had been given the year before, only his allowance was raised to 10 *momme* of silver per day.

Hereupon, on May 14th, 1801, Inō left Yedo, accompanied by Gunzō Hirayama, Sōhei Hirayama, Shūzō Inō, Keisuke Ogata, and a servant, and proceeding westward, opened a survey of the coast. Beginning with the western coast of Yedo Bay, he surveyed around the Miura Peninsula, and along the coast of Sagami and the eastern coast of Idzu, reaching Shimoda in about 6 weeks. The eastern coast of Idzu was so precipitous that passing along the shore was often impossible; sometimes, he stretched the measuring-rope on the sea from a boat, and at other times, he climbed steep rocks in order to take azimuths. This cost him endless pains—especially the transportation of bulky instruments like the large magnetic theodolite. As he found that the western coast was still more difficult of passage, he sent back the large magnetic theodolite to Yedo from Shimoda, and using the middle-sized magnetic theodlite in its place, surveyed the western coast in about 2 weeks. When he reached Mishima he received the newly constructed perambulator sent from Yedo, which he tested in surveying the Tōkai Highway eastward, and on July 16th again reached Yedo. Leaving Yedo again on the 29th of the same month, he made surveys up the eastern coast of Yedo Bay, then round the Bōsō Peninsula, and along the coast of Hidachi until after 8 weeks he arrived at a place in Mutsu called Karasuzaki, where he planned to make an observation of the lunar eclipse of September 22nd. Cloudiness, however, defeated his purpose. Continuing the survey of the coast for a week, he reached Matsushima. The coast between Yedo and Matsushima was comparatively even, and not much indented with capes and bays so that little difficulty in surveying was encountered. The coast north of Matsushima as far as Kamaishi and Miyako was much indented, and, to make matters worse, the coast grew precipitous enough to prevent passing along the shore. Therefore, after leaving Matsushima, he frequently surveyed the coast by the measuring-rope stretched on the sea from a boat until he arrived at Kamaishi on November 1st; and still proceeding north, he went round the Tonami Peninsula, and through Noheji and Aomori, and reached Mimmaya on Dec. 8th. Having thus completed a survey of the coast-line as he had planned, he started back again. Re-surveying the Ōshū Highway in spite of the severe cold and heavy snow, he reached Yedo on Jan. 10th, 1802.

For this journey of Inō's the Shogunate only issued a written order to the village officials, requesting that men and horses be provided for Inō at the fixed prices. However, the signatures of the Commissioner of Communication and the Commissioner of Finance, and the highly dignified form of the message made the village officials assist Inō in the matter of lodgings, fording, and obtaining information as to the route of his survey, and furnish helpers in surveying as well as carriers for the measuring-instruments. This practically meant much more help for Inō than that rendered the year before. As the way of meeting the expenditure was, however, the same as before, the sum Inō had paid out of his own pocket in defraying the necessary expenses amounted to more than 60 *ryō* (about ¥ 800) in addition to the allowance granted by the Shogunate. Moreover, as Inō surveyed the distances along the coast with a measuring-rope in order to secure greater accuracy, he could not proceed so fast as in the preceding year. The distance he surveyed a day averaging only 10 km., he divided his men into two parties, which worked separately. As his men grew skilful in their work, he left all common survey-work to them, while he himself was engaged in measuring the azimuths of distant mountains, capes, etc., and in nightly observation of the heavenly bodies, which were to serve as a control in map-making. Thus he cleverly divided the labour between himself and his men to secure the more rapid progress of the enterprise. He succeeded in making a fairly accurate survey of the coast-line. On the highway between Mimmaya and Sendai on his return, he was hindered by snow-storms and could merely test his survey of the preceding year by pacing again the distance. Nor did he use the measuring-rope between Sendai and

Yedo, but surveyed about 35 km. every day with the perambulator.

As soon as Inō reached Yedo, he worked with his pupils on the reduction of the survey-materials and upon map-construction. Though the survey of this year was not yet to be considered exceedingly accurate, the computation on the survey-materials resulted in figuring 28·2 *ri* (110·8 km.) as the length of one meridian degree—a value exactly equal to that adopted as his final result, after many an accurate survey, in his later years. The maps were completed in three scales in April, 1802. The large map is plotted on a scale of 1:36000 and consists of 32 sheets. The middle-sized is plotted on a scale of 1:216000 and consists of 2 sheets; and the small one is plotted on one sheet, on a scale of one half of that of the middle-sized. In the middle-sized and small maps the region surveyed in 1800 is also given. Of these maps, the large and the small were presented to the Shogunate, and the middle-sized to the Junior Secretary.

(*D*) *The Survey of the Coast of Dewa and Echigo*

The surveying and mapping of the eastern coast of the Main Island being completed, Inō's next task was naturally the construction of an accurate map of Yezo, as part of his own programme as well as from the standpoint of the defence of the country. It is evident that he and Takahashi again set about obtaining assistance necessary for a survey of Yezo. Though the time was not mature for the granting of his request for a survey of Yezo, his high technical skill in surveying and mapping was recognized by the Shogunate authorities through the maps he had presented. Since Inō's original idea had been, after all, the construction of a survey-map not only of Yezo, but also of all Japan, he came to consider it more sensible and expedient, instead of obstinately insisting on an accurate survey of Yezo alone, to persuade the authorities to consent to his further survey of the coast of the

Main Island with greater assistance and consideration than formerly. After consulting Takahashi, Inō decided to survey first the coast of the Japan Sea from Mimmaya in Mutsu to Echizen, and then the Pacific coast from Suruga to Owari, thus completing survey of the coast-line of the eastern half of Japan. Takahashi submitted this proposal to the Shogunate authorities and also made great efforts to secure more assistance and consideration from them. The Government consented to the plan, and in July, 1802, issued orders for Inō to survey the above-mentioned districts. And Inō was authorized during this survey to requisition free of charge a certain number of men and horses (5 men, 3 horses, and additional men to carry a long chest), and given an allowance just sufficient to defray the expenses of the journey including lodgings. He was to travel like a man on official business, paying at the low rate of "shelter" or at the price of "the rice eaten and fuel needed." His treatment was bettered and assistance increased, as compared with the preceding years.

As orders were given to start, Inō, intending first to survey the north-western coast of Mutsu, the coast of Dewa and Echigo, and also the intermediate highways during that year, left Yedo on July 10th, 1802, accompanied by Gunzō Hirayama, Shūzō Inō, Keisuke Ogata, Yūsuke Ōhira, and two servants. The party proceeded north along the Ōshū Highway surveying it once more accurately and reached Shirakawa in 10 days; then surveying the Dewa Highway, after passing Wakamatsu, Yonezawa, Yamagata, and Kubota (Akita), they reached Noshiro Port on the coast of Dewa on Aug. 20th. Inō had planned to make an observation of a solar eclipse on Aug. 28th, and awaited the day with elaborate preparations. But the day was cloudy and the phenomenon was barely visible through clouds so that an accurate observation was impossible. On Aug. 31st, the party left Noshiro and surveyed the highway through Hirosaki to Aomori, and then the coast-line—minutely surveying the coast from Aomori to Mimmaya which they had covered the year before, and then turning south down the Japan Sea coast of Mutsu, they arrived at Noshiro again on Sept. 25th. Still continuing survey of the coast of Dewa, the party went around the Oga Peninsula and passing through Tsuchizaki, Sakata, etc., entered the province of Echigo, and by way of Niigata and Kashiwazaki reached Imamachi (Naoyetsu) on Oct. 30th. From this point the party started on their return, surveying the Echigo Highway through Zenkōji (Nagano) and Uyeda as far as Oiwake, wherefrom they took the Nakasen Highway, and by way of Karuizawa, Takasaki, and Kumagaya, the party reached Yedo again on Nov. 18th.

For the surveying of this year, the Shogunate had not only instructed the village officials, in the name of the Commissioner of Communication and the Commissioner of Finance, to furnish Inō free of charge with a certain number of men and horses for carrying the measuring-instruments, but also it had issued a message to the clans on the route, requesting them to give Inō facilities for his work. So Inō received more help than before in his efforts to perfect the accuracy of his survey. Moreover, as Takahashi doubted Inō's evaluation of the length of one meridian degree which he had reached as the result of his survey-work of the year before, he was determined this time to test his value by improving the measuring-instruments and enforcing more strictness in the method of surveying. As for the main route whether along a coast or a highway, he always accurately surveyed the distances with an iron-chain, or a rattan-rope which he daily standardized by an iron-chain; in determining the direction of the route, he confirmed it by a double-survey in order to eliminate errors; an inclination of the route formerly determined by eye-estimation was now measured each time with a simple quadrant. At the chief points he stopped for several days to determine the latitude by repeated astronomical observations. Though the accuracy of survey was considerably increased, the skill of his men, attained in practise, and the operation in two groups in most cases, made it possible for Inō to survey a distance of from 12 to 25 km. a day. On returning to Yedo, he reduced all the materials of his accurate survey of this year, and therewith computed the length of one meridian degree. To his great relief, he got exactly the same value that he had obtained the year before. The map-draft of the surveyed districts was completed in February, 1803, but as the districts were only part of the area which Inō was commissioned to survey, he simply submitted the draft to the Shogunate authorities for inspection, and made preparations to start for the districts he had yet to survey.

(*E*) *The Survey of the Remaining Coast East of Owari and Echizen*

On April 16th, 1803, Inō left Yedo, accompanied by Gunzō Hirayama, Shūzo Inō, Keisuke Ogata, Ōye Muratsu, Ryōsuke Ono, and two others, for the purpose of surveying the remaining coast-line east of Owari and Echizen. The party proceeded west, surveying the Tōkai Highway from Yedo to Numadzu. There they commenced a survey of the coast-line along the provinces of Suruga, Tōtōmi, Mikawa, and Owari, and also a survey of smaller highways branching off from the coast-line to various castle-towns such as Fuchū (Shidzuoka), Hamamatsu, Yoshida (Toyohashi), and Nagoya; and on June 30th, they arrived at Saya, situated on the boundary between Owari and Ise. Leaving the coast there, the party surveyed a highway running through Ōgaki, Sekigahara, and Kinomoto, and reached Tsuruga in about two weeks. Here they began to survey the Japan Sea coast, first going west down the coast of Echizen as far as the boundary of Wakasa, and then coming east along the coast-line of Echizen and Kaga and surveying small highways branching from the coast-line to the castle-towns of Fukui and Kanazawa. On Aug. 21st the party arrived at a village called Imahama in the province of Noto. Here Inō, for the first time, divided the work among his men:—Gunzō Hirayama who was most experienced in survey-work leading a party completely independent from the main body surveyed clockwise the coast of the Noto Peninsula, and Inō led the main body in surveying the coast of the peninsula counter-clockwise. The two parties finished the coast of the peninsula in about 3 weeks, when they, again united, undertook a survey of the coast of Etchū and Echigo. On Sept. 27th, the party arrived at Imamachi where their work now joined their last station of the year before. Thus a survey of the coast-line of the eastern half of the Main Island was completed. Then the party sailed to the island of Sado whence, a survey having been made, they sailed back to Teradomari on Nov. 1st. From this point, they surveyed the highway by way of Nagaoka and the Shimidzu Pass as far as Takasaki where they connected their work with the preceding year's station and they were back in Yedo on Nov. 20th.

The work of this year was only a continuation of the preceding year's work, and the protection and treatment Inō received from the Shogunate was exactly similar. His method was also almost the same as that of the year before, only by adding a semicircular magnetic theodolite, Inō attempted to replace the large theodolite which had been so difficult of transportation. The village officials along the route were not only asked to help the party with the transportation of the instruments, lodging, ferrying, etc., but also to erect bamboo-poles in advance, to be used in sighting so that survey-work might be accelerated; and also to hand in a list of the names of the counties, villages, and feudal lords along the route, as well as the size of the fief and the number of the houses, etc. so that these might be referred to in the map-making. From the time Inō had commenced his survey-work in 1800, he had gradually improved his methods of surveying and his measuring-instruments, so that these were almost perfected by this time. During the journey this year, however, many members of the party were attacked by illness, to Inō's great annoyance. While surveying the Pacific coast, Inō left the common survey-work to his men, he himself chiefly working on the azimuth measurement of distant mountains, and measuring the latitude at each place where they stopped for the night. He thus succeeded in dividing the labour and accelerating the work. In Ōmi district, however, one of his men was attacked by measles, one after another all but Inō, even including the servants, being infected with it. When they reached Tsuruga, only Inō and one other man were able to work on the surveying, much to his embarrassment. Fortunately the others recovered before long and, dividing themselves into two parties as before, they worked assiduously. In Noto, Hirayama performed splendid service as the head of an independent party. In spite of the considerable delay due to illness, they succeeded in surveying the scheduled area, as a whole, at about the same rate as the year before.

Four years had elapsed since Inō had commenced a survey of Yezo, and he had already travelled several thousand miles. It cannot be denied that he had met with some slight hindrances during this time. His modest demeanour, however, with which he always observed had generally kept him in peaceful and favourable relations with the village officials and people along the route. Now, for the first time, a very unfortunate affair occurred. While surveying the coast of the Japan Sea, he came to Itoigawa in the province of Echigo. The officials of the place deceived him often with false statements in order to save themselves trouble, and caused him considerable embarrassment in the process of his survey. He scolded them in anger, but as they offered every apology, he did not carry it further. In order to conceal their own fault, however, the officials reported the affair, accusing Inō falsely, to the feudal lord, who, regarding the surveyor guilty, communicated this to the Shogunate before Inō's return to Yedo. Thereupon, the authorities suspecting Inō's abuse of his authority and overbearing conduct while travelling, ordered Takahashi to send Inō a dispatch of reprimand. On his return, of course, it was made clear that the accusation was false, and Inō was not further disturbed. This was the first and last instance in which he was treated by village officials with unprovoked malice and that he was threatened by any such danger.

The close of his survey in 1803 meant the completion of the survey of all the coast-line of the eastern half of the Main Island. On his return to Yedo, therefore, he not only planned to map the districts he had covered that year, but also desired to reduce and harmonize all the materials he had collected in the four years since his first survey of Yezo, with a view to constructing a complete map of the coast of the eastern half of Japan. Assisted by his pupils, he spent his whole time on this project.

But before this map of the coast of the eastern half of the country had been completed, a misfortune more disastrous than any hitherto met befell Inō. It was the death of Yoshitoki Takahashi, his beloved master, on Feb. 15th, 1804. Takahashi had not only instructed Inō in the new western astronomy, used his influence in helping to realize the survey-work Inō had proposed, but also imparted to Inō all the new knowledge he had gained in his tireless scientific researches. Now that Inō had completed the survey of the regions planned as a first stage and had set about constructing a map covering all the regions surveyed, much was expected of Takahashi, who would have been of great help to Inō, because of his profound scholarship and great originality, in the reduction of the materials and in the work of map-projection.

His death occurred within only three months of Inō's return. Inō's sense of loss may be imagined, though he no doubt found comfort in the thought that, while his master lived, he had determined the magnitude of the earth, a problem in which Takahashi had taken keen interest. It is a matter of deep regret that, as will be discussed later, owing to Takahashi's death, his idea of map-projection was not thoroughly realized and consequently unexpected defects appeared in Inō's map.

The map of the coast of the eastern half of Japan was completed about a half-year after Takahashi's death, and on Sept. 4th, 1804, was submitted to the Shogunate by Kageyasu Takahashi, son of Yoshitoki Takahashi, and Hidemasa Yoshida, colleague of Yoshitoki Takahashi. The map was drawn in three scales. The large one is plotted on a scale of 1:36000 and comprises 69 sheets for the Main Island alone. In this large map, the surveyed route is minutely drawn with a fine red line, even to the most trifling curves. Besides the boundaries of the provinces and counties along the route, the locations and names of villages, towns, temples, and shrines, and the ownership of the land, whether belonging to the Shogunate directly, to a Daimio, or a shrine or a temple, and the names of the Daimios, are minutely written in. Not only the location and shape of mountains, rivers, capes, and bays are indicated, but the natural features of land are given in different colours. The lines indicating longitude, latitude, or the azimuth lines of the mountains are not drawn in this map. The middle-sized one is plotted on a scale of 1:216000 and consists of 3 sheets, including Yezo. The surveyed route is likewise drawn in red, but the detailed features are omitted. The boundaries of provinces and counties, the names of towns, harbours, and villages are written in, but neither the class of territories nor the names of the Daimios. The notation of the topography is rough. As for the meridians and parallels they are drawn by Sanson and Flamsteed's system taking the meridian of Kuroye-chō in Yedo as standard; the chief azimuth lines of important mountains and islands sighted from various points are marked in red, and the values of the azimuths are also written in; and on the margin of the map is given a list of the latitudes measured at about 300 important points and of the surveyed distances between these points. The small map is plotted on a scale of 1:432000, and all the regions surveyed are drawn on one sheet. The notation of the surveyed routes, meridians and parallels, azimuth lines, mountains, rivers, towns, villages, and other features are marked in the same manner as in the middle-sized map, though some are abridged. Instead of the list of the latitudes and distances, this map has an introduction by Inō and prefaces by Kageyasu Takahashi and Hidemasa Yoshida. The introduction by Inō gives the gist of the methods he adopted in his survey and map-making, and the length of one meridian degree. In his preface, Kageyasu Takahashi writes that his father, sometime before he died, had an opportunity to read the Astronomia; and according to the size of the earth given in the work, he computed the length of one meridian degree for a mean latitude of 38° between Idzu and Mutsu; and to his great joy, his evaluation coincided with the one Inō had obtained by his survey; that Inō paid much attention to making a fine magnetic compass and succeeded in getting one with no instrumental error, and when he tested its pointing, it did not deviate to the west as indicated in western books, but pointed true north; and that the magnetic compass Inō made and the scale he used in his survey and mapping are being presented with the maps to the Shogunate for future reference. Yoshida's preface is to the same effect as Takahashi's.

Although the Shogunate authorities had already recognized Inō's ability in surveying by the maps of the south-eastern coast of Yezo and of the eastern coast of the Main Island, which Inō had presented to them, these maps were not quite adequate for general use, since they only represented localities along the coast-line. When the map of the coast of the eastern half of Japan was submitted to them, which he had prepared through co-ordination of all the materials of his survey, the true shape of eastern Japan was visualized before their eyes, so that the topography and facilities of communication on land and sea were to be seen at a glance; and as the advantage of a survey-map was thus given concrete proof, those who had been accustomed to mere sketch-maps were astonished at the rare skill Inō had exhibited. The map was spread out and pieced together in the great hall in the Yedo Castle for the inspection of the Secretaries and the Junior Secretary and other dignitaries; and the Shogun Iyenari, on hearing of the accuracy of the map, inspected it in person.

(*F*) *His Appointment as an Official and His Commission to Survey Western Japan*

As stated before, Shigetomi Hazama, when he went to Nagasaki to determine the longitude astronomically by observing the solar eclipse of Aug. 28th, 1802, on his way thither, attempted to obtain materials to determine the longitude geodetically by surveying the route between Ōsaka and Nagasaki. In the following year he consulted Takahashi, and secured permission from the Shogunate to survey the Kwansai region. If this plan had been carried out by Hazama, Inō's geodetic area would have been limited to the eastern half of Japan. Before the commencement of his work in the spring of 1803, however, Hazama was compelled to put it off till the following year, owing to some special hindrance. Meanwhile Takahashi's death and his dying wish brought Hazama back to the Astronomical Bureau at Yedo. This made his further survey impossible.

Inō's map of the coast of the eastern half of Japan was at this time submitted to the Shogunate, and it won high praise. In recognition of his services, the authorities appointed him an official on Oct. 13th, 1804, that he might develop his technical skill—an official called *kobushingumi* with a salary of "the rations for ten men," and was attached to the Astronomical Bureau as assistant to Kageyasu Takahashi, Official Astronomer. The position to which Inō was now appointed was only a humble one, but it was a rare privilege for a commoner to be raised to even such a post. It must have been a great joy to Inō who had always been eager to elevate his family name. Under such circumstances the survey-work which Hazama himself was to undertake was naturally transferred to Inō. The opportunity now presented itself to give scope to his ability throughout the whole Empire. The death of his master, though a heavy blow to Inō, nevertheless, brought him this good fortune.

On his appointment to this official position, Inō eagerly worked on a plan to carry out a survey of the western half of Japan and submitted his programme to the Government. On Jan. 25th, 1805, the Junior Secretary issued orders to Inō, through Kageyasu Takahashi, to make a survey of the whole of western Japan. According to this schedule, he was to survey the Tōkai Highway as far as Kuwana; to undertake a coast-survey—to go round along the coast of Ise, Shima, Kii, and Idzumi as far as Ōsaka; to proceed by way of Kyōto and survey the circumference of Lake Biwa; to go up to Tsuruga, and then to survey counter-clockwise the coast of the Chūgoku region, including that of the adjoining islands; to sail by way of Awaji Island to Shikoku to survey the southern coast; to sail to Kyūshū to survey the coast clockwise, including the adjoining islands; to sail back to Shikoku to survey the northern coast; to sail to Ōsaka again and survey the Tōkai Highway as far as Kuwana; to survey the highway through Nagoya and Iida to Kōfu, and the Kōshū Highway connecting Kōfu and Yedo; and then to return to Yedo. To save the trouble of returning to Yedo for every New Year season as before, this journey was to last continuously for 33 months in order to complete the survey of the area.

For this reason, the number of men for survey-work was to be increased considerably from that of the previous eastern journey. Moreover, since Inō was now to fulfil an official task in the capacity of an official immediately connected with the Shogunate, fairly qualified substitutes for him had to be provided, lest some circumstance should prevent his carrying out the work himself. Besides his pupils, therefore, two subordinates of the Astronomical Bureau, and Zensuke Takahashi (who was later called Kagesuke Sukezaemon Shibukawa), the second son of Inō's former teacher, were added to his party. As to the manner of obtaining men and horses and the payment of expenses, there was also here considerable improvement, with the change of Inō's status. Inō was empowered to engage, free of charge, a certain number of men and horses for carrying the measuring-instruments and baggage, and a dignified message to this effect was sent to village officials from the Commissioner of Communication and the Commissioner of Finance. To the lords of the clans along the route, the Government issued orders to render all necessary assistance to the party in fording, lodgment, and any other matter necessary for their survey-work. As survey-work became an official affair of the Shogunate and the clans began to regard it seriously, the authorities gave instructions to Inō and his subordinates that they should be especially prudent in speech and behaviour while travelling, that they should neither abuse their authority nor exaggerate the importance of their work to the country officials and people, and that they should use their best endeavour to accomplish an accurate survey. To his pupils joining the party, Inō repeated the instructions, and each had to deposit a written oath. Now, the carriage of the measuring-instruments, the ferriage, the expense for repairing the coast-road, etc., that is, the greater part of the expenses of the survey-work was defrayed by the local authorities and the people along the route. What the party had to pay while travelling was only a very small sum for lodgment, the fixed rate for a traveller on official business, and some incidental expenses. However, Inō and his subordinates were granted a fair allowance, in addition to their regular travelling pay; and writing-brushes, paper, ink, candles, etc. were all supplied to them. As compared with the time when Inō surveyed the eastern region, this was a far better financial arrangement.

(*G*) *The Survey of the Chūgoku Region*

After several busy weeks in which he selected the personnel of his party, constructed more measuring-instruments for his larger company, and made other preparations for the journey, Inō started out on his project to survey western Japan on March 25th, 1805. Those who accompanied him were Zensuke Takahashi, Kinsuke Ichino, a subordinate of the Astronomical Bureau, Teibei Sakabe, the same, Gunzō Hirayama, Shūzō Inō, Seijirō Montani, Tōjirō Nagasawa, Kwampei Kosaka, all pupils, and several servants—fourteen men in all. They commenced survey-work from Takanawa in Yedo. Nearly every day they divided themselves in two parties; and each party surveying the Tōkai Highway westward at the rate of from 6 to 8 km. a day; they reached Maisaka in about 3 weeks. After spending about 10 days in surveying the coast of Lake Hamana, they proceeded along the Tōkai Highway until they arrived at Kuwana on May 7th. Then they undertook a survey of the coast. Covering the coast of Ise and the highway running along it, they reached

Yamada after more than 10 days and here sojourned for several days, making a survey of the coast-line in the vicinity, in order to prepare for the observation of the occultation of one of Jupiter's satellites, on the night of May 20th. The observation was newly arranged this year to determine astronomically the longitude of a survey-station, and this was the very first that was accomplished while travelling. The party then went to Toba where about 10 days were spent in surveying the coast in the vicinity; and after again observing an occultation of one of Jupiter's satellites, they surveyed the southern coast of Shima and Ise, in two or three parties, and on July 10th arrived at Nishikiura on the eastern boundary of Kii. There they stopped and made preparations to observe the lunar eclipse at dawn of July 12th. As the sky was clear, fortunately, they fulfilled the purpose of the observation. When they now began the survey of the coast of Kii, the indentation was so great and the sea so rough that they could not make much progress. By the time they finally reached Shingū, they were several months behind their scheduled date. Their goal was still so far away and no one could tell when their task would be completed. Inō found the men in extremely low spirits. Doing his best to encourage them, he continued to work, dividing the men into two parties. At length they reached Wakayama on Sept. 1st; and surveying the smooth coast of Idzumi, they reached Osaka on the 10th of the same month. Ichino who had been unable to participate in the work because of illness and was not yet recovered and Montani for another reason left the party here and returned to Yedo. The other members stayed about 10 days in Ōsaka while surveying in the vicinity, then proceeded along the Yodo River as far as Kyōto; and after surveying the city and its suburbs, went to Ōtsu on Oct. 5th and surveyed the circumference of Lake Biwa in about 5 weeks.

The original plan had been to go to Tsuruga after completing the lake and survey the Japan Sea coast west of Wakasa. Owing to the delay, however, it was now the most inconvenient season for surveying this district. So the plan was changed to that of first surveying the coast of the Inland Sea. The party, now recruited by Masagorō Shimokōbe, a subordinate official who had arrived from Yedo to take the place of Ichino, left Ōtsu on Nov. 13th and surveyed the course of the Uji River as far as Fushimi, where they measured the circumference of Ōike (the Great Pond). The coast of Settsu west of Ōsaka was surveyed, then that of Harima and Bizen; the surveying of the adjoining islands taking considerable time, Okayama was reached on Jan. 20th, 1806. It had been Inō's original plan to assign one or two members to the examination and preliminary reduction of the survey-materials each day, but the shortage of hands had prevented this from being put into effect. It was now imperative that they stop at a suitable place long enough to study and reduce the materials they had gathered lest they should be confused beyond measure when the time came for plotting. Therefore, he spent about a month and a half (including the New Year of the Japanese calendar) in Okayama in reducing the survey-materials, in preparing drafts for mapping and in astronomical observations, especially in that of the occultations of Jupiter's satellites. At the same time he repeatedly negotiated with the Shogunate authorities through Takahashi in Yedo for an increase in his staff, in view of the impossibility of carrying out the survey-work as planned, because of the unforeseen difficulties along the coast of western Japan. He succeeded in obtaining an addition of two men to his party and it was arranged that the party should return to Yedo for a time, after surveying the coast of the Chūgoku region.

They left Okayama on March 7th, 1806 and, dividing themselves into three or four parties each day, surveyed in a westerly direction the coast of Bitchū and Bingo, and adjoining islands.

As they entered the province of Aki, Keisuke Ogata and Hayata Kadokura from Yedo joined them. Continuing their survey-work, they came to Hiroshima on May 17th and to Akamagaseki (Shimonoseki) on June 22nd. Thus, they had surveyed the coast of the Chūgoku region along the Inland Sea. Because Inō had been suffering from ague since about the middle of the month, he was treated for some time at Akamagaseki. He was not quite restored to health, but in spite of his illness he started from Akamagaseki with the others for a survey of the coast along the Japan Sea. He gave himself solely to recruiting his health, leaving all survey-work to the men under him, until he completely recovered. After leaving Akamagaseki, the party began to survey the north-western coast of Nagato, which they found entirely different in its topography from the coast of the Inland Sea, less indented and with few adjoining islands. Because of this, their progress was rapid and Hagi was reached after ten days. Still continuing their work, on Aug. 2nd, they came to Matsuye, after having surveyed the coast of Iwami and Idzumo. As they were ready to sail from Mihogaseki for Oki Island, Inō's illness took a sudden turn for the worse and he came back to Matsuye to rest. The party sailed with a favourable wind and landed at the island on Aug. 17th; and after surveying the whole island in about two weeks, they sailed back to Mihogaseki. Having surveyed the circumference of Lake Shinji, they came to Matsuye on Sept. 15th to meet Inō. At that time Inō was feeling much better and so he left Matsuye with his men to survey the coast of Hōki and Inaba. They reached Miyadzu in Tango on Oct. 16th. It seems that Inō was at last completely restored to health, after four long months since the first attack of the disease. From this place they still continued surveying the coast, getting to Obama in Wakasa about a half-month later. Here they divided themselves into two parties, one of which surveyed the coast-line as far as the boundary of Echizen and connected with a survey-station of 1803 while the other party surveyed the highway from Obama to Tsuruga where the two parties met. Here they parted again and one surveyed the highway running from Tsuruga along the eastern coast of Lake Biwa while the other the highway along the western coast of the lake. The two parties met at Ōtsu on Nov. 28th, whence Zensuke Takahashi left for Ōsaka on account of his private concerns and the others started on their return journey to Yedo. Surveying, on the way, the Tōkai Highway between Ōtsu and Yokkaichi, the Sangū Highway between Seki and Tsu and the highway between Kuwana and Nagoya, they finally reached Yedo on Dec. 24th.

During this journey, now that Inō was a full-fledged Shogunate official, the clans along the route sent out their retainers to receive him and instructed village officials to do their duty faithfully; when the party surveyed the islands in the Inland Sea, boats chartered by the clan were at their service so that they might not feel any inconvenience in performing their duty; and when carrying their measuring-instruments over precipitous mountains, several times the fixed number of men and horses (six men, one horse, men for carrying a long chest, besides one man and five horses for carrying the baggage of the members of the party) were offered for accelerating their work; and in other preparations and assistance for survey-work, the order of the surveyors was very faithfully observed. This being the case, their work should have progressed steadily this time, but what with Inō's lack of accurate knowledge of the topography of the Kii Peninsula and the conditions of the archipelagoes of the Inland Sea, what with the frequent illness of Inō and his companions, the original plan could not be carried through. The party was compelled to leave its work unfinished and return to Yedo in spite of Inō's original programme to finish it at a stretch. The accuracy of their survey was more and more increased; an attempt was made to determine longitude by observing occultations of Jupiter's satellites; a sketch of the scenery of the mountains, islands, rivers, and seas along the route was undertaken by one of the members in order to illustrate the topography. Moreover, literature and maps for reference were demanded of the local officials and people. Thus the materials for mapping were abundant.

A most unfortunate affair occurred during this journey. As already stated, subordinate officials of the Astronomical Bureau were added to the party as its staff; but Hirayama, one of Inō's pupils who had been trained as a surveyor for years and was now an expert, was sometimes disposed to slight these officials and gradually an estrangement arose between them. Moreover, while Inō was unable to superintend the enterprise owing to his long illness, discipline became lax among the younger pupils; they made merry at the hotels, coveted paintings and curios, and often scandalized the people along the route by their rough manners. As everything was known to the Government, the authorities, upon the return of the party to Yedo, demanded that some punishment be given the pupils. It ended in the expulsion of Gunzō Hirayama, the ring-leader of the pupils who had opposed the officials and Kwampei Kosaka, the chief violator of discipline. This journey was the most eventful that Inō ever had experienced and it was the only instance throughout his career in which because of the lack of his leadership what might be termed a failure ensued.

On his return to Yedo, Inō used his own house at Kuroye-chō in Fukagawa as the office for map-construction. Here Zensuke Takahashi, Sakabe, Shimokōbe, and the pupils who had accompanied him on the journey worked on map-making, a fair allowance being all granted by the Government. At the same time they carried out some fundamental observations in order to determine the declinations of fixed stars. In about a year, maps of three kinds were completed according to the former scales. These were submitted to the authorities in January, 1808.

(*H*) *The Survey of Shikoku*

Inō, having learned many lessons from his survey of the Chūgoku region, especially the disadvantage of travelling for a long period, decided to survey the remaining region in two separate trips—one for Shikoku and another for Kyūshū. For the purpose of surveying the coast of Shikoku and a few highways on the way, Inō left Yedo on February 21st, 1808. His party consisted of Teibei Sakabe, Denzaemon Shibayama, Masagorō Shimokōbe and Katsujirō Aoki, all subordinate officials, Shūzō Inō, a pupil, Bunsuke Uyeda and Saemon Kuboki, the same, and servants;—sixteen all told. The composition of the party was exactly similar to that of the Chūgoku party.

On their way from Yedo to Shikoku, they surveyed the Kiga Highway running from Hamamatsu through Kiga to Goyu and a few other smaller highways. They sailed from Maiko to Awaji Island and commencing a coast-survey of the island on March 31st, from the northern end, measured the eastern and southern coasts for about 10 days, until they came to Naruto Cape, whence they sailed for Muya in the province of Awa. There a survey of the coast of Shikoku was begun. First going south along the eastern coast of Awa, then entering Tosa and by way of Muroto Cape, they arrived at Kōchi on May 25th. A detachment was sent from here to survey the highway across the province to the boundary of Iyo, while the main body proceeded still westward along the coast. The coast west of Kōchi, however, became more and more indented with capes and bays, and less and less headway was made. Uwajima in Iyo was at length reached on Aug. 12th. Still proceeding along the coast and around Sata Cape, the party came to Matsuyama on Sept. 30th, whence a survey of the northern coast of Iyo and adjoining islands was made. They reached Kawanoye on Oct. 26th. Here a detachment was sent to measure the highway to the boundary of Tosa, to connect with the survey-station of the highway measurement previously taken from Kōchi. After spending about two weeks in surveying the coast of Sanuki they reached Marugame; and while there, surveying the Shiaku Archipelago, they made a splendid observation of the solar eclipse of Nov. 18th at Tomariura of Motojima. After finishing the archipelago, they continued the coast-survey of Sanuki and also that of Shōdo Island and other islands. Muya, the terminus of the survey of Shikoku was reached on Dec. 24th and their work on the island was now completed. The party then sailed again to Awaji to survey the western coast, to complete the survey of the whole island. This done, they took ship to Hyōgo and went to Osaka. Here Shūzō Inō left the party to return to Yedo on account of illness while the others started from Ōsaka on Jan. 11th, 1809, for a survey of the highway running through Nara, other noted places in Yamato, and Uyeno in Iga, to Rokkenjaya in Ise, thus completing their plan for this journey. It was on New Year's Day by the Japanese calendar, when they visited the Great National Shrine and they reached Yedo again on March 3rd.

The assistance the party received on this journey from the officials and people along the route was exactly like that on their previous journey. The feudal lords showed more and more respect to Inō and the highways along the coast were often repaired, or even newly built for the convenience of his survey-work. Each clan also sent, besides the ordinary helpers in surveying, several retainers to accompany the party and be at their service. And usually some officials followed Inō under pretense of being apprentice-students of land-surveying. This fact shows, on the one hand, the strict observance on the part of the feudal loads of the Shogunate orders, but also, on the other hand, their precaution regarding survey-work. It may be said that while Inō enjoyed much assistance from the feudal lords, his movements were constantly under observation. His strict handling of his men during this journey prevented any unfortunate affair from taking place. His work was accomplished as he had planned it.

After returning to Yedo, Inō was engaged in map-making as usual, with the subordinate officials and his pupils who had accompanied him. In August a large map consisting of 22 sheets, a middle-sized map of one sheet, and a small map of one sheet, and *Fūkeidzu* (The Description) of the route in 11 volumes were completed and submitted to the Shogunate. At this time the Government, which was needing a reliable general map of the whole country, ordered Kageyasu Takahashi to construct it. At his request Inō made the map on a scale of 1:864000, using his materials for the regions he had already surveyed and temporarily drawing the regions he had not surveyed by referring to old maps. This was submitted to the Shogunate with his other maps above-mentioned.

(*I*) *The First Survey of Kyūshū*

As soon as the mapping of the Shikoku region was completed, Inō made preparations for a survey of the Kyūshū region and left Yedo on Oct. 6th, 1809. He was accompanied by Teibei Sakabe, chief assistant, Masagorō Shimokōbe, a subordinate official, Katsujirō Aoki and Jinzaemon Nagai, the same; Eizō Yanada, a pupil, Bunsuke Uyeda and Ryōsuke Hakoda, the same, and nine servants. Divided in two parties, they surveyed the highway of Iwatsuki as far as Kumagaya, where they commenced an accurate survey of the Nakasen Highway westward. On the way, they made an observation of the lunar eclipse of Oct. 23rd, at Matsuida, and measured the circumference of Lake Suwa.

The survey of the Nakasen Highway was finished in about two months after leaving Yedo. Now surveying the Yamasaki Highway from Yodo in the province of Yamashiro to Nishinomiya in Settsu and thence the Sanyō Highway, the party reached Akamagaseki (Shimonoseki) on Jan. 29th, 1810. Previous to this, Inō had frequently attempted to observe the occultations of Jupiter's satellites, but each time had been more or less prevented by cloud or mist. He stopped several days at Akamagaseki, especially for this observation, but in vain. He then took boat to Kyūshū.

This being the New Year season according to the Japanese calendar, they stopped some two weeks at Kokura. The coast-survey of Kyūshū was then undertaken. On Feb. 15th, they commenced the survey eastward along the coast of Buzen and Bungo. Though an attempt was made to observe the solar eclipse of April 4th at Hatoura in Bungo, the sky was so overcast that the phenomenon was barely observed amidst clouds. They continued surveying the coast, but as the eastern shore of Bungo was much indented with capes and bays, they were delayed in their work. Nobeoka in Hyūga was reached on May 8th. Going south down the unindented coast of Hyūga, they reached Obi on the 29th of the month. Sometimes dividing into two parties and again uniting, they surveyed the coast of Ōsumi, the highway connecting Obi with Fukuyama, by way of the Ushi Pass and Miyakonojō, and the small highway across the province of Ōsumi, and reached Kagoshima on July 24th. During their ten days' sojourn there, they were engaged in a survey of the island called Sakurajima, in obtaining an accurate measurement of latitude, and in the observation of the occultation of one of Jupiter's satellites. Surveying the coast of Satsuma, they came to Ichikiminato, whence they went by boat to Koshiki Island, and after surveying the whole island, were back by boat near Ichikiminato on Sept.

17th. Inō's original plan had been to proceed to Yakushima and Tanegashima, directly from Kagoshima, but the stormy sea of the season compelled him to survey the Higo district first. Here Inō divided the party in two, himself leading the main body while the other was placed under Sakabe's command. Sometimes divided and at other times united again, they surveyed the coast from Ichikiminato to Yatsushiro in Higo, also the Amakusa Islands, the highway running from Ichikiminato across the province of Satsuma to Kagoshima, the Sappi Highway connecting Kagoshima with Yatsushiro through Hitoyoshi, and some other smaller highways. They were at Yatsushiro on Dec. 13th. The illness of some members at that time caused considerable inconvenience to their survey-work and consequently the voyage to the two islands in the south of Satsuma was postponed until their next tour. The party, therefore, continued to survey the coast of Higo to the boundary of Chikugo, where a measurement of the highways was undertaken. After surveying the highway from Sekimachi through Yamaga to Kumamoto, they arrived at Kumamoto on Jan. 3rd, 1811. The highway connecting Kumamoto and Ōita across Higo and Bungo was now measured and the New Year (according to the Japanese calendar) was celebrated at Ōita. Then surveying the highway connecting Ōita with Kokura, the party reached Kokura on Feb. 9th. Here they temporarily discontinued the survey of Kyūshū and started on their return journey.

On the way home, in two parties or again in one, they surveyed the highway between Akamagaseki and Hagi, that connecting Ogōri with Ōmori in Iwami Province through Yamaguchi and Hamada, that between Hamada and Hiroshima, that between Hiroshima and Miyoshi, that between Ōmori and Fukuyama by way of Miyoshi, that between Miyoshi and Okayama across the province of Bitchū, and some others. The party arrived at Okayama on April 21st. Then the highway between Himeji and Arima, and some others were surveyed. They now went to Mino, Owari, and Mikawa to measure the small highways in these provinces; and surveying the highway from Okazaki to Suwa through Iida and some adjoining highways, the party came to Kōfu and, measuring the Kōshū Highway from there, reached Yedo again on June 28th.

During this journey Inō met with no hindrance worth mentioning. The relations with the feudal lords were friendly; though he apparently expected some difficulty in a powerful clan like Kagoshima, he was there given handsome treatment and assistance and because of this his survey-work was accomplished quite smoothly. The topography of the coast of Kyūshū, however, was so much more rugged than Inō had imagined that his plan could not be carried out. Consequently the opportunity to sail for Yakushima and Tanegashima was lost when their plan was changed half-way, so as to leave out some area for the next survey-tour and to return to Yedo to reduce their survey-materials and make maps. The method and form of surveying were similar to those of the previous journey; only they laid great stress on the observation of the occultations of Jupiter's satellites. This astronomical observation proved almost fruitless, but the survey of a number of highways connecting many important points along various routes, which was carried out in the Tōkai, Tōsan, and Chūgoku districts on both the outward and homeward journeys, proved to be valuable control-materials later on, in constructing the complete map of Japan and were also helpful in providing greater detail.

Inō stayed a half-year in Yedo, and was engaged with his subordinates and pupils in mapping. Towards the end of 1811, the maps for the district he had this time surveyed were completed in all three sizes and submitted to the Government authorities.

During this period Rinzō Mamiya returned from Yezo, and being ordered to make an accurate survey of the far north, lived with Inō in order to become more practised in land-surveying, especially in latitude measurement.

(*J*) *The Second Survey of Kyūshū*

On Jan. 9th, 1812, Inō left Yedo on a journey to survey Kyūshū for the second time. Teibei Sakabe, chief assistant, Jinzaemon Nagai, a subordinate official, Matabei Imaidzumi and Seijirō Montani, the same, Keisuke Ogata, a pupil of Inō, Ryōsuke Hakoda and Keizō Hoki, the same, and eleven servants accompanied Inō on this trip. The party spent about three weeks on the outward trip in surveying the route running from Fujisawa, a stage of the Tōkai Highway, to Kōfu through Gotemba, and the highway coming back from Kōfu to Okitsu on the Tōkai Highway along the Fuji River. With no more measuring on the way, they arrived on March 8th at Kokura at the northern end of Kyūshū. While the sea was calm in the spring, they desired to survey Yakushima and Tanegashima, the two islands whose survey had been put off on their previous trip. So the party at once undertook a survey of the following highways towards Kagoshima:—the highway connecting Kokura with Yatsushiro by way of Kurume, Sedaka, and Kumamoto and then running near the coast from Yatsushiro to Ichikiminato, the one between Yamaga and Ōtsu through Waifu and another one between Yunoura and Nakano through Ōguchi. They arrived at Kagoshima on April 14th.

While the ships for crossing to the two islands were being prepared, the party stayed in Kagoshima, surveying the streets of the city and making astronomical observations. The local clan fearing the danger of the voyage and the inconvenience of survey-work, chartered eight large boats for the party; and a large number of helpers in surveying as well as carriers were to be sent with the party from the mainland. About a week was necessary for such preparations, after which they set sail. Having no favourable wind, however, they were stopped again for two weeks at the Yamakawa Harbour, the southern end of Satsuma. Yakushima was finally reached on May 8th. The survey of the island was completed in a little over ten days, but they had to stay two more weeks, waiting for a favourable wind. At last they succeeded in getting to Tanegashima on June 5th. This island was also measured in a little more than ten days. After waiting another two weeks for a favourable wind, they came back to Kagoshima on July 1st. About ten weeks had elapsed between Inō's departure from Kagoshima and his return. The real survey-work of the two islands was finished in only three weeks, all the rest of the time being wasted in awaiting a favourable wind. Inō had originally planned, making the most of this opportunity, to survey Iwōjima, Kuchinoerabujima, and the Seven Islands to the south of Satsuma. Hearing the difficulties in executing the plan from the clan officials with whom he negotiated, and anticipating a waste of many days on the voyage, he gave it up, and satisfied himself with merely determining by distant sighting the position of the islands within sight.

The party, now divided in two, Sakabe being the leader of the second group, and then united again, surveyed the following highways:—the one running from Hamanoichi in Ōsumi to Takanabe by way of Kirishima and across Hyūga Province; that from Takanabe along the coast of Hyūga and through Mimitsu to Nobeoka; that running from Nobeoka across the northern part of Hyūga along the Gokase River and through Takamori, connecting with the Oita Highway at Sakanashi; that connecting Hitoyoshi in Higo Province with Sadowara across Hyūga Province; that running from Mimitsu in Hyūga through the mountain to Mamihara in Higo and then to Kumamoto; that from Uchinomaki in Higo to Kumamachi and Morimachi in Bungo and then to Yabakei; and the highway between Yabakei and Kumamachi and the path to Mt. Ehiko. The party assembled again at Kokura for the observation of the lunar eclipse of Aug. 22nd. The sky being clear on that night, they succeeded in observing the phenomenon satisfactorily.

Again they engaged themselves in survey-work, proceeding west from Kokura along the coast of Chikuzen as far as Hakata, and measuring the adjoining islands. As they entered Hizen Province and went west, they found the coast more and more indented, till they often adopted a cross-traverse in order to control the coast-survey. On Oct. 17th, the party arrived at the boundary between the clan of Saga and that of Hirado, or the boundary between Saga and Nagasaki Prefectures of to-day. They now undertook a survey of the following highways, sometimes in two parties and sometimes in one body:—the highway between Imari and Saga; that from Saga to Yamaga; that from Yamaga to Kumamachi; that running from Kumamachi to Yanagawa through Kurume; and other smaller highways. This required about a month. Then they took to a coast-survey again, starting their work at the boundary between Chikugo and Higo. Surveying the coast of Chikugo and Hizen and some highways near the coast, they arrived at Isahaya on December 5th; and then continued their work round and across the Shimabara Peninsula and came to Yagami where, leaving the shore of the open sea, they crossed the isthmus to the coast of Ōmura Bay and reached Ōmura on Dec. 26th. The party under Inō proceeded northward, surveying the eastern coast of Ōmura Bay and the highways near by, while the party under Sakabe surveyed the western coast of the bay as far as the northern end of the Sonoki Peninsula. It being near the end of the year (by the

Japanese calendar), on Jan. 30th, 1813, the two parties met at a small village near Saseho called Shidzuura and celebrated the New Year. Inō stayed there to observe the lunar eclipse of Feb. 15th, surveying the vicinity, but observation was impossible because of the cloudy night. On the following day the party left Shidzuura and, surveying the coast and adjoining islands, reached Hirado in about two weeks. After completing a survey of Hirado Island in about half a month, Inō proceeded eastward along the northern coast of Hizen to the clan boundary of Saga, and thus connected up with the previous survey made from Kokura.

The party, after surveying a few islands near by, sailed to Iki Island with a favourable wind on April 13th. This island was completed in about ten days; and after a few days' waiting for a fair wind, they sailed to Tsushima. This island, situated in the extreme north-west of the Empire, is exceedingly precipitous along most of its coast. As its position had never hitherto been studied, Inō surveyed its coast-line most carefully while at the same time trying to determine its relative position to the continent by frequent distant sighting of the azimuths of the various high mountains of Korea, and took special care in measuring latitude and observing the occultations of Jupiter's satellites. His survey of the island was finished on June 14th. With a fair wind, after a few days' waiting, they sailed to Ukujima, one of the Gotō Archipelago. Sakabe's branch co-operating with Inō's main party in surveying the archipelago, on July 26th, they reached Fukuye in the island of the same name at the sonthwestern end of the archipelago. A few days previous to this, Sakabe had been taken ill, while on Hinoshima, and had preceded the others to Fukuye for rest. Since he was not quite himself yet, Inō left him at Fukuye to recuperate, and Inō and the others were engaged in surveying the island. On Aug. 8th, however, a report reached the surveying-party telling them of the dangerous condition of the patient. Inō and the others hurried back to the hotel at Fukuye and did their best in nursing the sick, but on the afternoon of August 10th, Sakabe breathed his last on this remore island in the far west. Inō collected all the members of Fukuye and, in great sorrow, buried the dead in the grounds of a temple named Sōnenji. They gave up their work in mourning for a week, and at the same time considered how to cope with the situation in the face of this unforeseen misfortune.

Inō decided to put Jinzaemon Nagai in charge of the branch, when working in two parties. Fukuye Island was finished on Aug. 24th and, surveying several islands on their voyage back to the mainland of Kyūshū, they returned to the Sonoki Peninsula on Sept. 2nd. The party then started a coast-survey at the northern end of the peninsula and, going south down the western coast, reached Nagasaki in about ten days. During the half-month they stayed here, they were engaged in surveying in the vicinity and making astronomical observations. On Sept. 26th, the party resumed the coast-survey from Nagasaki and proceeding for about two weeks round the headland of Nomo, connected at Yagami with the last station of the previous year's coast-survey made from the direction of the Shimabara Peninsula. The whole coast-line of Kyūshū had now been surveyed.

They now took to highway-surveying. In two parties, or united, they surveyed the following highways:—the highway between Yagami and Togitsu; that from Sonoki to Shioda; that between Kawatana and Takeo; that from Ogi to Hakata; that between Hakata and Ashiya; that from Akidzuki to Kokura through Kōbaru; and other smaller highways connecting with these. They arrived at Kokura on Nov. 3rd, having completed the survey of Kyūshū.

The party now started up on its return journey. On the

way they surveyed the following highways:—the highway running from Odzuki in Nagato across the province to Shōmyōichi on the northern coast and then to Hagi; that from Tsuwano in Iwami to Tsuda in Aki and then to Hatsukaichi; that from Miyaichi in Suwō to Tsuda in Aki; that running from Tsuda through the mountains in Aki and Iwami to the Ōmori silver-mine; that running from Hiroshima through the mountains of Bingo and Idzumo to Kitsuki; that from Onsendzu in Iwami to Tottori by way of Matsuye, Yonago, and Kurayoshi; that from Yonago to Himeji through Tsuyama; that between Kurayoshi and Kuze; that between Tsuyama and Okayama; that from Tottori to Himeji through Chidzu; and many other smaller highways connecting with these. They were at Himeji on Feb. 18th, 1814, where they celebrated the New Year (by the Japanese calendar). They then resumed their work and surveyed the following highways:—the highway running from Himeji across Harima and Tajima to Kinosaki; that from Himeji to Kyōto through Kaibara and Sasayama in Tamba; that between Toyooka in Tajima and Kyōto through Fukuchiyama and Sonobe in Tamba; and the highways connecting many other important points. The party arrived at Kyōto on April 16th. They now surveyed the following highways:—the highway between Kyōto and Nara; that from Nara to Seki through Uyeno in Iga; some highways in Owari and Mino; the highway running from Gifu to Takayama in Hida and then turning to Yabuhara; that from Semba to Shinonoi through Matsumoto; that between Zenkōji (Nagano) and Iiyama; that from Oiwake (the forked-road) of the Nakasen Highway to Shimonida and then to Honjō; and that running from Kumagaya to Chichibu-Ōmiya and then to Itabashi by way of Kawagoe. The party reached Yedo again on July 9th, 1814.

As already stated, Inō's method of land-surveying had been

fairly arranged in 1803, but as he surveyed western Japan as a Shogunate official, his survey-work grew more and more minute and the items of his investigation collateral with surveying more and more detailed. Especially during his survey of Kyūshū, he was anxious to make his coast-survey in the strict sense worthy of the name. No piece of land by the sea, however rugged and precipitous, was left unmeasured, if humanly possible. Not one of the numberless islands scattered like checkers escaped his attempt to investigate its topography. The natural features along the route were carefully sketched for reference in plotting maps. So, if the treatment of the members of the survey-party was improved, they had to labour harder. Besides, since the astronomical determination of the longitude by observing the occultations of Jupiter's satellites was part of their work, Inō and his men, already exhausted with the day's survey-work, had frequently to spend much time and labour at night. Though the party had the misfortune to be deprived of its adjutant by illness, any other member suffered no serious impairment of health during this second survey of Kyūshū, which lasted two and a half years. Their relations with the feudal lords along the route were always favourable and much assistance was given by these. This made it possible for them to carry on their work steadily in the western islands, in spite of dangers on land and sea. A survey of more highways than planned on both outward and homeward journeys naturally prolonged their journey. As a whole, however, Inō's mission was fulfilled according to programme.

At the time of his second trip to Kyūshū, Inō was already advanced in years and, aiming at a speedy success, was sometimes too rash, so that, Teibei Sakabe assisted him by sharing and managing his affairs; Keisuke Ogata, his pupil, ever keeping company, did his best to comfort Inō's mind and body. Besides, perhaps because of Inō's assiduous work at such an age, the feudal lords gave him a warm reception everywhere he went; and men, hearing of his great reputation, came one after another, asking to be admitted as his private pupils. This must have given Inō inward satisfaction. The latter half of the journey, however, was to bring him pain because of a series of unforeseen events. In May, 1813, while engaged in a survey of Tsushima, the sad report reached him of a fire which had taken place at Takahashi's residence in the Astronomical Bureau, Asakusa, Yedo, in March that year and had reduced to ashes a number of important documents, including observation records of the occultations of Jupiter's satellites taken at the bureau which were to be used together with Inō's observations in computing the longitudes of different points. On their way from Tsushima to the Gotō Archipelago they received the sad news of the death of Sakabe's mother. Then came the great misfortune in August, when Sakabe himself succumbed to illness on Fukuye Island. Sakabe had a position at the Astronomical Bureau because of his proficiency in Japanese mathematics and was a subordinate under Kageyasu Takahashi. Ever since Inō had undertaken the survey of western Japan, he had accompanied and assisted Inō on his tours. He was naturally a good-tempered man, and as he faithfully applied himself to the business for years, none in the party could compare with him in experience as well as in technical skill. Sakabe was indispensable to Inō as his right-hand. Robust as he was, and in the prime of life, never having injured his health in spite of exposure to the roughest weather for more than eight years, he fell a prey to a disease at the age of forty-two towards the completion of the great enterprise. Inō gave way to grief. In his letter to Kagetaka, his eldest son, reporting this sad event, he said, "Since he has been my wings, I am disappointed and sorrow-stricken, I assure you, like a bird that has lost its wings." But who would have thought that Kagetaka, to whom he thus reported his sorrow, appealing for sympathy, had already died of illness on July 4th? Fearing the result of this blow to the father in the far west, the family kept the matter secret as long as they could. It was more than two months before the sad news was finally sent to the father. It must have been about the time Inō completed the survey of Kyūshū that the bad news reached him. A great pity that such calamities should have befallen him in succession! Inō's perseverance, however, did not allow him to give himself over to excessive grief so as to injure his health, but led him to devote himself more than ever to the completion of his task. Having fulfilled his mission, Inō now returned to Yedo.

His residence at Kuroye-chō in Fukagawa, having proved so small as to cause much inconvenience in map-making, Inō had asked his friends to seek a more suitable house for him, but a desirable one had not been found as yet. Recently Kuwabara, one of his relatives, had moved out of a residence on Kamejima-chō, Hatchōbori, near the Nihon-bashi, and Inō decided to take this place. On his return from Kyūshū, he had the residence repaired and on July 20th, 1814, moved into this new residence on Kamejima-chō, leaving the old one at Kuroye-chō, in Fukagawa, which, since his coming to Yedo in 1795, had been the base station of his astronomical as well as geodetic measurements and the headquarters of the survey of the whole country. The site was located before the earthquake of 1923 as the lot at the back of houses from No. 15 to No. 17, Itchōme, Kamejima-chō, Nihonbashi Ward, its area covering some 500 square meters. This new residence became Inō's new mapping-office where he, with his subordinate officials and the pupils who had accompanied him on the journey, was engaged in reducing the survey-materials and map-construction. Meanwhile a chain-survey of the highways within Yedo was undertaken. The reason was that in starting a survey in various directions Inō roughly measured the highways within Yedo simply by pacing, or by the perambulator, and made it a rule to commence an accurate survey from Takanawa-Ōkido, Yotsuya-Ōkido, and Senjū-Ōhashi, or the entrances and exists of Yedo; but now that the surveys in various directions had been completed, he needed an accurate survey of the distance between the Nihon-bashi—the origin of the distance-measurements—and these entrances and exits of Yedo which had been used as temporary origins. The measurement of the highways between the Nihon-bashi and these doors to Yedo was made in March, 1815. Though the record of this measurement is missing, the survey of a few roads of some 10 km. each was no doubt completed within a few days by a small number of surveyors.

(*K*) *The Survey of the Seven Islands of Idzu*

Inō had a desire to survey the islands scattered in the south of Idzu, or the so-called Idzu-shichitō (Seven Islands of Idzu), to complete the survey-map of Japan and also to revise the old measurements by re-surveying the eastern coast of the Idzu Peninsula. He had negotiated with the Government authorities in the summer of 1814 and obtained an unofficial consent. When he finished the chain-survey of the highways within Yedo in the spring of the following year, Inō was given formal orders from the Shogunate to proceed to the survey of the Seven Islands of Idzu, the coast of Idzu, and some highways adjoining.

At this time Inō was already seventy years old and his health was somewhat affected. After his return from the second survey of Kyūshū, he was also more or less in low spirits. His daughter and Kageyasu Takahashi, entertaining fears as to his ability to brave the rough sea, surveying the Seven Islands in person, prevailed upon him to give up the voyage and send his subordinate officials and pupils to survey the appointed region, Inō himself superintending their work from his home in Yedo.

Thereupon, a survey-party was organized composed of Jinzaemon Nagai, Yaoji Sakabe, heir of the late Teibei Sakabe, Seijirō Montani, subordinate officials, Sadadū (Ryōsuke) Hakoda, one of Inō's pupils, Keizō Hoki, the same, and servants—eleven men, all told. The party under the leadership of Nagai left Yedo on June 4th, 1815. They went to Mishima in Idzu without making any survey and then commencing a survey of highways, reached Shimoda in about a week. While waiting at Shimoda for a favourable wind to sail to the Seven Islands, they surveyed the vicinity and measured the mountains and islands by distant sighting. Leaving the Shimoda Harbour aboard a ship chartered by the Government on June 25th, the party sailed for Hachijōjima, but as the wind changed, they reached Miyakejima first. They landed and surveyed part of the island. With a fair wind they sailed on June 28th for Hachijōjima, reaching it on the following day. A long rain had set in; but as one of their chief purposes was to make an observation of the solar eclipse of July 7th on the island in order to determine its longitude astronomically, a station was chosen in a small village situated in the eastern part of the island, where, as the rain stopped, the whole party went and prepared themselves for the observation. Unfortunately, on that day the sky was overcast and the eclipse took place behind a cloud, defeating the purpose of their observation. Then the party commenced to survey the island, first its coast, then Kojima, an island belonging to it, and then its highways. The operation was completed on July 22nd. Now that the observation of the solar eclipse had failed, the only means left them in determining the longitude of Hachijōjima was to do so geodetically by measuring the azimuths of the various islands through distant sighting and to connect them gradually with the mainland. So the party climbed high mountains and tried to measure the azimuths of the various islands, until July

31st when their task was finished. They set sail for Miyakejima with a fair wind on Aug. 4th from Hachijōjima. A dead calm and a rapid current on the way caused them to drift in three days to Misaki in Sagami. There they landed to plot their map-drafts and to sight distant mountains and islands, while waiting for a fair wind. They were at last able to sail to Miyakejima on Aug. 15th. Making the island their base, they wished first to survey Mikurajima. Boarding four fishing-boats laden with surveying-instruments and provisions, a large party of seventy including the crews, after braving a rough sea, reached Mikurajima. Partly because of the cliffs all around and partly because of the dangerous waves, much difficulty was experienced in their survey-work. By climbing down a precipice on a rope, by swimming with a measuring-rope from rock to rock, by other feats of adventure, and by enduring hardships, the party finished the survey of the whole island in ten days and returned to Miyakejima. About a week was then spent in surveying this island and two weeks in waiting for a favourable wind to sail for Kōdzushima. On Sept. 14th, they arrived at Kōdzushima which they surveyed in about a week. Since both this island and Mikurajima lie in a position most favourable for determining the position of Hachijōjima, they were determined to sight it. For this purpose they prepared a lookout on the top of a mountain on the island, to which the members were despatched in turn for sighting Hachijōjima as soon as fair weather permitted it. Their purpose was finally fulfilled on Oct. 6th. Then after a week they sailed to Niishima and in about three weeks this island and two other islands, Shikinejima and Jinaishima had been surveyed. They now returned to Susaki in Idzu with a fair wind and surveyed the coast and examined their drafts while awaiting a good wind. On Nov. 21st, they sailed to Ōshima, which they surveyed in about a fortnight. Their commission to survey the Seven Islands of Idzu having been thus fulfilled, the party sailed back to the Shimoda Harbour on Dec. 10th.

They now prepared for the observation of the total lunar eclipse of Dec. 16th at Shimoda. The weather being fair, they were able to make a satisfactory observation. So they departed from Shimoda on Dec. 20th, and spent about a month in surveying the eastern coast of Idzu and a few short-cutting highways. As it was now the end of the year (by the Japanese calendar), they all went to Atami to celebrate the New Year and stayed there about a month, arranging the survey-materials and preparing map-drafts. Leavng Atami on Feb. 22nd, 1816, the party undertook the survey of the following highways:—the highway between Mishima and Gotemba; that between Gotemba and Yoshiwara; that connecting Yoshiwara with Hitoana by way of Ōmiya; that running from Gotemba through the Otome Pass to Yumoto; the circumference of Lake Ashi; the highway running from Hiratsuka through Atsugi to Hachiōji; the highway from Atsugi to Yedo; that running from Hachiōji through Kawagoe to Kumagaya; the route along the Arakawa from Kumagaya to Senjū; and some other minor highways. The party arrived in Yedo on May 8th, having fulfilled its mission.

During the journey of surveying in the Idzu region, though Inō did not go with the party, the members kept on good terms with one another, preserved good health, enjoyed pleasant relations with the local officials and people, and carried out their work very peacefully. The principle of survey was similar to that of the preceding year, but the dangerous places and rough sea which characterize the islands of Idzu, the great distance of Hachijōjima from any other island, and the remarkable local disturbance of terrestrial magnetism on the island caused much difficulty and often required special care. The efforts of the members won them success in the face of these many difficulties.

On their return to Yedo, the members were engaged in drawing maps of the district they had surveyed. It is not clear when they completed the maps, but because they had reduced the survey-materials while waiting for wind, and also because they had sojourned at Atami about a month to prepare map-drafts, the maps must have been completed in a comparatively short time.

(*L*) *The Survey of the Streets of Yedo*

What Inō did in Yedo during the period in 1815–1816 when his subordinates were surveying the Idzu region, is not clear. However, it is evident that following his presentation to the authorities of the maps of Kyūshū after his second survey, he received orders to make a map of the whole of Japan by uniting and revising the maps that he had already prepared. It is possible that he was making preparations for this map. The western part of Japan was surveyed after Inō had been appointed an official of the Shogunate and because of the various facilities he was given and the amount of time he expended, not only the coast-line, but also the highways, the courses of the rivers and the circumferences of the ponds and lakes were minutely surveyed. On the other hand, the survey of the eastern part had been made by him as a ronin or a lordless samurai and in a very brief time; and on that account, the only highways he surveyed were several main ones; nor did he survey Kasumigaura, the Tone River, or other rivers, lakes, and ponds. For this reason he was anxious, before drawing the map of the whole country, to make a supplementary survey of the eastern part, following the survey of the Idzu region, so that the minuteness of the two parts might be unified. At the beginning of 1816, he submitted his plan for such a supplementary survey to the Shogunate and negotiated with the authorities. After much delay Inō's proposal was finally rejected about August of the same year. Nevertheless, the chain-survey map of the highways within Yedo, which Inō had drawn after surveying some of the main roads connecting the Nihonbashi with the entrances of Yedo, had so interested the authorities that they now intended to order Inō to make an accurate survey of the streets of Yedo and plot a detailed map. Before long Inō was formally commissioned to survey the streets of Yedo. Setting to work on Sept. 29th, 1816, he completed the survey on Dec. 1st of the same year. Those who were engaged in the task were, besides Inō, Masagorō Shimokōbe, Matabei Imaidzumi, Jinzaemon Nagai, Yaoji Sakabe, Seijirō Montani, Keijirō Watanabe (formerly Keisuke Ogata), all subordinate officials and some pupils and servants. The map based on this survey was drawn on a scale of 1:6000, and presented to the Government in the following year, 1817.

(M) Rinzō Mamiya's Exploration of Saghalien and Eastern Tartary and His Survey of Yezo

As already stated, the region Inō had desired to survey before any other was Yezo. Since making merely a rough survey of the south-eastern coast of the island in the first year of his geodetic career, he had been obliged to postpone the survey of the whole coast owing to the rejection of his demand for necessary preparation; and meanwhile he had surveyed the eastern coast of the Main Island, circumstances compelling him to extend the survey of the eastern coast to that of the eastern half of the country and again to undertake the survey of the western half. He had reached the advanced age of seventy without resuming the unfinished task in the northern island which had been his first object. Had the unrivalled fame as a land-surveyor which he now enjoyed led him to neglect the survey of Yezo? Had the strain of the foreign relations which had prompted the survey of Yezo as a pressing need in national defence been alleviated?

No, Inō had not neglected his original object. The foreign relations had spurred him on with still greater urgency to define the topography of the northern frontier. The reason why he had applied himself, not in this direction, but in the survey of Japan proper was that one of his pupils, a most competent explorer, had been active in Yezo, Saghalien, and Manchuria, steadily performing his mission.

This explorer, whose name was Rinzō Mamiya, was born in a lonesome village in Hidachi Province, in 1775. Born in a very poor family, he received no adequate education in his childhood. However, he was naturally bright and gifted with mathematical ability. When seventeen years old, he was made a humble official employé of the Shogunate; and his obtaining in 1800 a position connected with Yezo affairs led him to investigate its frontiers. He first went about the islands, such as Kunajiri, Etrofu, and Shikotan, and investigated the customs, the natural features, and the configuration of the capes and bays. In this his talent was exhibited, though he had not much knowledge of land-surveying, nor did he carry suitable measuring-instruments. He made a rough survey more or less like a sketch, with some rudimentary knowledge of land-surveying and a crude magnetic compass. His chance meeting with Inō this year in Yezo led him to form his acquaintance and finally to study under him; his more profound knowledge on the subject was later imparted by Inō. After this he was generally held by his work in Yezo. When the Russians plundered Etrofu Island in 1807, he chanced to be there and his presence of mind and opportune action in the very jaws of death won him great praise. Though he was then engaged in exploring and surveying in the island, it seems that he had not yet begun a regular undertaking like the coast-survey of Yezo, which proved so helpful to Inō in later years.

Previous to this, the Shogunate and Matsumae Clan had felt

the necessity of studying the topography and conditions in Saghalien, which lies to the north of Yezo Island and had sent men there frequently for inspection. Tokunai Mogami, for instance, had already explored it in 1792 at the request of the Shogunate, and much information had been obtained concernng the state of things in southern Saghalien. Nothing whatever was known as yet of the northern part. There was even a divergence of opinion as to whether Saghalien was an island or a peninsula. This being the case, the Commissioner of Yezo Affairs, in 1808, appointed Mamiya an official of the Shogunate because of his proficiency in foreign exploration and ordered him to inspect Saghalien in company with Denjūrō Matsuda. So the two men sailed to Shiranushi, the south-western corner of Saghalien. It was arranged that Mamiya should inspect the eastern coast and Matsuda the western. On May 12th, Mamiya sailed from Shiranushi northward up the eastern coast reaching the Kita-Shiretoko Cape on June 14th. Thence he turned back to Manue and crossing the isthmus to Kusunnai on the western coast, he sailed north and overtook Matsuda. Then they turned round the Nakko Cape on July 15th; and having explored as far as the estuary of the Rakka, they started together on their return. After about a month they reached Sōya in Yezo Island.

As a result of this exploration, Mamiya and Matsuda learned the topography and the general conditions of the southern part of Saghalien and almost ascertained that Saghalien was not connected with the continent, but was an independent island. However, Mamiya was by no means satisfied with this exploration and was anxious to produce better results. Another permission having been granted, Sept. 3rd of the same year saw Mamiya resolutely leaving Sōya alone for the north. Sailing from Shiranushi northward up the western coast of Saghalien, he reached a place called Tosshokau after braving many dangers and spending a numbe of days. Then, because the natives he had hired earnestly implored him to return south, and also because of want of provisions, he came back to Tonnai to go into winter quarters. Early in the spring of the next year, 1809, he sailed north again and arrived at Noteto on May 22nd; and Nanio on June 24th. He now personally perceived the existence of a strait between Saghalien and the continent and proved that Saghalien was an island. Now returning to the vicinity of Rakka, he crossed the strait to the continent. Travelling by land from Mushibō, he reached the Mangō River (the Amur) by way of Lake Kichii. He then explored the upper reaches of the river and came to Deren on Aug. 21st. Here he had an interview with a Chinese official. Going down the river and travelling south along Manchurian coast, he studied the relative position between the continent and Saghalien. Crossing the strait to a place in Saghalien called Wakee a little north of Rakka, he returned to Sōya on Nov. 5th.

Mamiya carefully wrote down his observations during his journeys. These he compiled, on coming home, under the title of the "Kita-Yezo-dzusetsu" (Northern Yezo Illustrated) and the "Tōdatsu-kikō" (Eastern Tartary Travels) and also drew maps of Saghalien and eastern Tartary based on his survey. In all these works he showed the topography, the customs, and the real conditions of these regions. The map drawn in 1810 on a scale of 1:36000 and consisting of 7 sheets is still preserved in the Department for Compiling Materials for History of the Tōkyō Imperial University, and a map drawn on a reduced scale of 1:864000 is also preserved in the Tōkyō Imperial Library. These maps were drawn by Inō's subordinates at the request of Kageyasu Takahashi. The latter map, though on a reduced scale, gives meridians and parallels, from which the general quality of the survey may be guessed. According to this, the configuration of the coast-line of Saghalien and Eastern Tartary which Mamiya roughly measured is almost correctly represented. But the measurement of distances is incorrect and gross errors may be detected in the latitudes of different points. This is to be accounted for by the fact that Mamiya did not carry any useful instruments for astronomical observation; that his knowledge of astronomical observation was then so extremely meagre that he was unable to use expedient methods for determining the latitudes; that the distances of different points were mostly determined by eye-sighting on board a boat; and that the topographical features were chiefly determined only by the azimuths of the more prominent objects measured with the help of the stick capped with a magnetic compass which he had obtained from Inō and which was so accurate and so easy in handling.

After his successful exploration of Saghalien and eastern Tartary, Mamiya was promoted to a higher position as a Shogunate official and his salary increased. He was living in Yedo in 1811 when he received orders to survey the northern frontier again. Just at the time Inō was back in Yedo from his first survey of Kyūshū and drawing maps. Mamiya went to live with Inō to practise such methods as the accurate determination of latitude by astronomical observation, and thus considerably increased his knowledge of land-surveying. Inspired by his high spirit, Inō instructed him earnestly. On their parting in January, 1812, when Inō was to proceed to Kyūshū again and Mamiya to the northern frontier, Inō penned a composition entitled "On the Departure of Rinzō Mamiya," in which he gave expression to his wish for his pupil's success. The manuscript of this composition in Inō's own handwriting is still preserved in the Inō family.

It is not to be doubted that Mamiya on his return to Yezo made use of the method of latitude measurement he had newly mastered in his work of surveying the island. We regret that no record of his survey has been preserved, to give us the details of the progress of his measurements. It is evident that he remained in the island till as late as 1815 or 1816 and that he was back in Yedo in 1817. And it must have been after he returned to Yedo that he gave Inō his survey-materials of Yezo.

Concerning his subsequent activities nothing is known, except that he was in the party of a local governor who inspected the Seven Islands of Idzu in 1828 and that he was several times sent to certain localities on some special secret service. In short, Mamiya contributed much in the realm of science through his positive adventures in his earlier life, while in his later life, he exerted himself in his negative position as a secret-service man and gradually departed from the field of science. He was responsible for the exposure of Kageyasu Takahashi's violation of a national interdict in giving a survey-map to Siebold. Although it was doubtless a right attitude to let no personal relations interfere in a public affair, this scandalized some people and gradually estranged those who had sought his acquaintance in admiration for his indomitable ambition. His last days were passed in solitude and in 1844 Mamiya died at the age of sixty-nine, in Yedo. Even Siebold, who was made to suffer considerably, though indirectly, by him, could not deny Mamiya's services in the field of science. In his work, "Nippon," Siebold writes as follows:—

"Aber zwei Männer waren es, deren Wissenschaftsliebe und Wissbegierde, deren Mut und Ausdauer nicht nur ihr Vaterland, sondern auch wir in Europa eine nähere Kenntniss von den Inseln zu verdanken haben, welche sich vom Norden von Japan bis nach der Mündung des Amurs hinziehen und von dem Lande selbst, welches dieser grosse Strom bei seiner Ergiessung ins Meer bespült. Sie heissen Mogami Tokunai und Mamia Rinso."

Mamiya's great services for our national defence and in the study of geography have since been recognized. In 1904 he was posthumously honoured with the Senior Grade of the Fifth Rank.

(*N*) *Inō Undertakes the Mapping of the Whole of Japan*

The survey of the streets of Yedo which Inō carried out in 1816 with his subordinates was his last field-work. From that time on he chiefly superintended his subordinates and pupils in mapping the whole of Japan by arranging and revising the survey-materials he had collected during the last seventeen years. When Inō set to work by Government order, he intended to complete the map by the end of 1817, at the latest. The work was much delayed, however, and at the end of 1817, he had to report that the work would need two more years before its completion. The reasons for this delay are not found in any record. But the chief explanation may be that the survey and mapping of the streets of Yedo after this drawing had been begun, engaged his men consecutively for a few months, and also that they arranged Mamiya's survey-materials and plotted the whole of Yezo for the purpose of putting it in the new map. Besides, some subordinates and pupils who were proficient in map-making, had to give up their work owing to illness or some other circumstances. Whenever a shortage of men occurred, Inō tried to fill up the vacancies. But this must have been one of the reasons for the delay. Above all, Keijirō Watanabe, one of Inō's oldest pupils, who was now engaged in the map-making as a subordinate official, was transferred to another post because of the death of his father-in-law. His departure from Inō's office affected the work, no doubt. Therefore, Inō implored the authorities to pardon Gunzō Hirayama, who had been excommunicated in 1806, and his great skill possibly increased the efficiency of the work of map-construction.

As for Inō himself, since his return to Yedo from Kyūshū and the establishing of his new residence at Kamejima-chō, he had some business, it is true, like the drawing of the maps of the

Kyūshū region and superintending the survey-party to the Idzu region, but it was by no means as taxing to his mind and body as a journey. Especially after 1816 or 1817, his chief business being only the supervision of the construction of the map of the whole of Japan, he now had leisure to take part in some affairs other than map-making.

While travelling about, a number of men became his pupils, but their relation to him was mostly nominal. Those who really desired to study astronomy or geodesy received his instruction by correspondence, chiefly after his return to Yedo, or came up to Yedo to receive his instruction directly. Inō carefully answered the questions of such eager students and assisted them in constructing or purchasing measuring-instruments and did much towards the popularization of the sciences. Also, about this time he began to think of compiling books for the benefit or enlightenment of the public. It was during this period that Inō, making use of the method designed by Chokuyen Ajima, wrote his "Taisūhyō-kigenjutsu narabini yōhō" (The Principles and Uses of Logarithms) in which he gives a fundamental table convenient to calculate the numerical value of common logarithms to the tenth decimal and at the same time revised the table of ten-place logarithms given in the Shu-li-ching-yün and also wrote the "Bukkoku-rekishōhen-sekibō" (In Confutation of the Indian Astronomy) in which he confuted "Bukkoku-rekishōhen," or the Indian Astronomy compiled by a Buddhist priest named Yentsū. Besides, his "Katsuyen-hassenhyō-kigenhō" (The Principles of Trigonomerical Tables) was no doubt prepared about this time.

The study of map-projection is extremely important in drawing the adequate map of an extensive area. Though Yoshitoki Takahashi had attempted to study it as early as 1802, Inō had not paid much attention to it as yet. In his maps of the coast of the eastern half of Japan drawn in 1804 he had ignored Takahashi's will. It seems that Inō now keenly felt the necessity of the study of projection as he undertook the map of the whole of Japan. One of the extant pamphlets in his own handwriting showing his investigation and computation on the subject may be regarded as a relic of this period. It is to be greatly regretted that the method described in the pamphlet was not adopted in drawing the map of the whole country.

(*O*) *Inō's Death and the Completion of the Map of the Whole Country*

Since his return from the survey of the Kyūshū region, Inō had become more and more infirm with age and had frequent attacks of illness. In 1815 or 1816, however, he had planned a supplementary survey of the eastern part of Japan to be conducted in person, if the authorities should approve it. After surveying the streets of Yedo in the autumn and winter of 1816, his health began to fail; in spite of increased nutrition, his chronic disease of moist cough attacked him oftener and oftener, and he also suffered from toothache. Until the early days of 1818, he found himself still able to supervise the map-construction and to instruct his students in the country by correspondence in answer to their questions. Before long his health became more precarious and he grew weaker and weaker. At last on May 17th, 1818, he died at seventy-three years of age in his residence on Kamejima-chō. When Inō became critically ill, he said, "My accomplishment of the great task of surveying Japan and my present reputation are due to the guidance and instruction of my teacher, Mr. Yoshitoki Takahashi. So I wish my remains to be buried beside the grave of my teacher, in token of the great debt I owe him." This dying wish of Inō's enhanced his name and added a brilliance to his personality, and it at the same time epitomizes the real truth of his great achievement. According to his last wish, he was laid by the side of Yoshitoki Takahashi in the graveyard of a temple named Genkūji, Kita-Kiyoshima-chō, Asakusa. The gravestones of these two great men standing side by side have always been devoutly looked up to as a monument to the profound affection existing between master and pupil.

Though Inō had died, his death was not formally announced as the map was not yet completed. Tadanori, his eldest grandson, being only twelve years of age, his eldest daughter lived at his residence in Yedo to bring up the boy and to take charge of the mapping office established at the residence. Though the construction of the complete map was still continued by the subordinates and pupils, its progress was naturally checked by Inō's death. Besides, Gunzō Hirayama was seized with illness and died in the winter of the following year; and Yaoji Sakabe also died some time later. In spite of the assistance of several subordinate officials of the Astronomical Bureau and especially that of Seiyen Kuboki, Inō's great friend, who frequently travelled between Tsunomiya in Shimōsa Province and Yedo and did much in making a fair copy of the survey-records and in other matters, the completion of the map was delayed considerably, and it was not finally finished until the summer of 1821. On Aug. 7th, Kageyasu Takahashi, accompanied by Tadanori Inō and the subordinates who had taken part in the drawing of the survey-maps, proceeded to the Yedo Castle with the large map, consisting of 214 sheets, the middle-sized map of 8 sheets, and the small map of 3 sheets, and the records of the coast-survey in 14 volumes. In the grand hall, the maps were pieced together in the presence of the Secretaries and the Junior Secretary, for their inspection, and the ceremony of presentation was completed. A copy of the middle-seized map was at the same time presented to Settsu-no-kami Hotta, Junior Secretary. The large map contained the whole of Japan from the two islands, Yakushima and Tanegashima, in the south-west to the island of Yezo in the north-east, on a scale of 1:36000. The method of plotting was similar to that of the large map of the eastern half of the country drawn in 1804. In the middle-sized map the whole country was drawn on a scale of 1:216000; and the method of its plotting was similar to that of the middle-sized map of 1804; the azimuths of the more important mountains and islands sighted at various points are given in red lines and their numerical values are also written in. As for the longitudes and latitudes, the meridian passing the site of the Calendaric Office at Nishi-Sanjō-dai in Kyōto is taken as the first meridian and meridians and parallels for each degree are drawn on Sanson and Flamsteed's system. The small map was drawn at a reduced scale of one half of the middle-sized map, with some omissions. "The Record of the Coast-Survey" mentions the latitudes of more than 1200 points on the coast and highways, the extension of the coast-line, the circumferences of the lakes and ponds, and the lengths of the highways which Inō surveyed; and, at the commencement, Kageyasu Takahashi writes a preface and Inō a preface and explanatory notes. Takahashi says in his preface, "Land-surveying would be impossible without a reign of peace. Without a proper man it would still be impossible. This has always been and still is by no means easy.... And behold these maps harmonizing heavenly measurement above and defining the earthly features below! Herein is reflected a flower of peace and civilization. And without Inō would this have been possible? In China with a history of fifty centuries, a map was made only in the Ch'ing dynasty and with the help of Westerners. No small services, these of Inō's." He fitly expresses Inō's great services to his country.

Although the work of map-making was over for the time being, with the completion and presentation of these maps, Masagorō Shimokōbe and the other members of the office were continually engaged in making duplicates of the various maps and also in drawing many copies at the informal request of the feudal lords. As the Shogunate, prompted by the necessity of survey-maps, made Inō extend his surveying further and further, so a number of the feudal lords came to realize this also in sympathy with the trend of the times. Therefore, the lords sought his favour by giving him a warm reception when he travelled in their territories, by utilizing those of their subjects who had become Inō's pupils, and by all other means. It is evident that the clans of Tokushima, Ōmura, Hirado, and Karatsu secretly negotiated with Inō to obtain the maps of their own territories or those of their neighbours. If it were possible to investigate thoroughly those who attempted to get the maps of the whole or sections of the country, the number would be quite large. Though the survey-maps were constructed at the request of the Shogunate and therefore not to be made public freely, it seems that the authorities permitted Inō and the others to some extent to make copies of the maps and distribute them. This is easily inferred from the fact that there was a rumour that the completion of the map of the whole country was delayed because Inō was busy in making copies for the feudal lords for his own profit, neglecting his official business. By the end of 1823, the duplicates of the maps of the whole country in three sizes, that of the city of Yedo and "The Record of the Coast-Survey" were all completed and handed over to Tadanori Inō for preservation. The task of the map-construction was now finally completed.

5 Inō's Character, His Family, and His Descendants

A man of tremendous fortitude, never yielding to difficulties, always ambitious, never relaxing in his efforts; and yet rich in common sense, polite in speech and always modest—such was Inō the man. These mark all his works, as hitherto narrated. The fact that, while travelling for years through the territories of powerful feudal lords who viewed him not without suspicion, he always sustained amicable relations with them and accomplished his purpose, is proof of his strong personality and unequalled energy. Concerning his family, however, many untruths have been circulated. We shall now inquire into the truth of the matter, as it will serve to throw light upon another side of his character.

The precepts which Inō himself wrote and which he gave his family will suggest his principles in these matters. "First, don't act falsely by any means," he wrote; "be faithful to your parents, brothers and sisters, masters and friends; be honest. Secondly, if advised or reproved by your inferiors as well as your superiors, listen to them and at all costs observe what they tell you. Thirdly, remember cordiality and modesty; be polite in speech and conduct; be humble in everything; never quarrel with others." However, a false story has been circulated concerning his domestic circumstances at the time he entered into the Inō family and this has caused a great misunderstanding before the public. According to this, he was first an errand-boy in the Inō family and the former master, taking a fancy to his talent, adopted him as his son-in-law. On this account his wife often despised him and adopted an overbearing attitude. While sitting at table one time with his wife, she, becoming displeased, scolded him angrily, "You don't deserve to be treated as a husband. Go out into the kitchen and eat with the servants." Inō appeared not to be offended by her remark and retired with the utmost calmness. This tradition, it seems, is a far-fetched story to impress people with Inō's magnanimity and indifference to trifles. Its incredibility can be easily understood from the utter untruth of the first half of the tradition. For Inō was not a servant in the family before he was adopted and the former master of the family, or the father of Inō's wife, had been dead some twenty years before their marriage. Besides, a study of his wife's character according to the original sources shows that she, being trained strictly by her mother, was more domestic even than the common run of women. Even if she were not able, like the wife of Yoshitoki Takahashi, to have sympathy with her husband's pursuit of studies, there is no doubt that she shared and co-operated with him in the task of replenishing the family property, which he regarded his first duty and did his best to realize. His home at this time was by no means destitute of taste or pleasure. His wife accompanied him on his extensive sight-seeing journey to Matsushima in Ōshū at a time when women's travelling was extremely difficult. Inō bitterly lamented over his wife's death and gave relief to the poor as a prayer for the repose of her soul. These acts should prove the existence of a tender affection between husband and wife.

Later he married a daughter of Kuwabara as his second wife, but he had the misfortune to lose her by death five years after their marriage. About the time he came up to Yedo to study astronomy seriously and to devote himself to the national enterprise, Inō's home was in a tragic state. For his eldest daughter whom he had most dearly loved, he had adopted Moriemon Furukawa, one of his relatives, as her husband, as she grew up and had established them in Yedo in the rice business. Now, Moriemon had somehow displeased Inō and created a disturbance in the family. Inō finally expelled Moriemon. His daughter who did not approve of this punishment went out with her husband against her own father. This caused Inō to disown her also and led to a tragic separation of father and daughter for years. But, on the other hand, Kagetaka, his eldest son, who had succeeded him, had thoroughly satisfied him in the point of observing his family precepts. This son was, after all, a man of mediocrity, with a character entirely different from that of his father and Inō could not expect much from him. Fortunately, Kagetaka's wife was a clever woman who adjusted the domestic affairs with him, and when Tadanori, his grandson, was born Inō began to enjoy some happiness in the family. In 1810, his eldest daughter returned home, obtaining pardon from Inō. She now pledged herself to make amends for her disobedience with a firm will with which she had gone back on her father rather than deny herself. And when she began to comfort him in his old age as well as she could, the deep sorrow that had depressed him for years now quickly disappeared. Harmony reigned again in the household.

Inō was not to enjoy this happiness very long, however. During his second journey to Kyūshū in 1813, Kagetaka died, leaving an infant. This brought loneliness afresh to his heart. Not long after his return to Yedo on the completion of his official business, another trouble arose in the family. From the first Inō appears to have intended to instruct Shūzō, his second son, in the art of land-surveying, because he took the son with him in the capacity of a pupil on his surveying-journeys, beginning with the trip to Yezo in 1800 when Shūzō was only fourteen years old and repeated this every year till the trip to Shikoku in 1808, letting him work on surveying and map-making. But when, according to the custom of a Shogunate official in those days, Inō deposited a written oath with regard to his wish in case of emergency, before leaving for the west in 1805, he neither nominated Shūzō as his possible successor, nor even mentioned him as one of his own sons. This was perhaps because he thought Shūzō unfit for the post, or perhaps because of some other unknown reason. After 1809, he did not take Shūzō with him on his surveying-journeys and about 1811 Shūzō was made to enter into a family named Sakurai. There existed no estrangement as yet between father and son; but as the imprudent son gradually took to heavy drinking and became reckless and loose in his morals, he was expelled from the Sakurai family about 1814. Dependent now upon his father, Shūzō grieved Inō bitterly. He never followed Inō's advice, but repeated follies and persistently opposed his will until the father disowned him in May, 1815. When previously Inō had disowned his eldest daughter, she could have claimed some reason on her side and asked for consideration. But the circumstances under which Inō now sent away this son were entirely different. No future repentance could be expected. At a time when his age was so advanced as seventy and his work of surveying and mapping needing more hands than ever before, what a pity to have had to drive out his own son who had been especially trained for these tasks!

Thus, Inō's family life which was normal in his earlier years was quite troubled as the years went by. He had no domestic happiness except in the thought that the two women, his eldest daughter and Kagetaka's widow, were ably conducting the family business at Sawara and that his eldest grandson, now studying with him, promised to develop rare talents in the future. It is rather strange that this man of noble character who succeeded as a business man, a village official, and a scholar should not have completely succeeded as an old master of a happy family. Was it that the cold tasteless home, in which he had been brought up after his mother's death in his childhood, had so affected his character in spite of himself, that his passionate affection for his children should have been suppressed by his rigid reason and that the nobleness of his character did not include a sympathetic understanding of his own children?

Such being the true state of his family in his last years, Inō, in his written oath with regard to the successor to his official position, which he handed in to the authorities, expressed his wish that his position as a Shogunate official be cancelled at his death.

that only his salary be handed down to Tadanori, his eldest grandson, and that the privilege to wear swords be extended to his descendants for ever. In the year Inō died, the mother of Tadanori, or Kagetaka's widow, also died, and consequently the business of the main family at Sawara had to be conducted for them by others. During the three years that Inō's death had been kept secret the family property was on the wane again. When the presentation of the maps of the whole country was over, the relatives assembled to discuss the question of Inō's successor, and they agreed to give up the position of Shogunate official. Then Inō's death was formally announced on Sept. 29th, 1821.

The Shogunate, recognizing Inō's conspicuous services and complying with his own wish stated in his written oath, declared that Tadanori should be allowed the rations of five men and be given a residence on Hakuya-chō in Yedo. It also extended to him and his descendants the privilege of wearing swords. The authorities also commanded Tadanori to master the art of astronomical observation and land-surveying and to engage himself in regular astronomical observations at his house in Sawara. That the Shogunate allowed Tadanori, not his grandfather's entire salary, but a half of it and gave him a new residence was in accordance with precedent, as Shigetomi Hazama had received about the same privilege for his services in compiling a new calendar in the Kwansei era. As Tadanori was still studying astronomy, it was decided that he should stay in Yedo and continue his study. In September, 1822, Tadanori's sole protectress, his aunt, Inō's eldest daughter, died and this caused another discussion as to his future. Some relatives, being persuaded by Masagorō Shimokōbe, formerly Inō's subordinate, and others, proposed that Tadanori should lead a party of his grandfather's pupils to survey the course of the Tone River, as well as the circumference of Kasumigaura and Kitaura which Inō had not surveyed, in order to repay the privilege granted him by the Shogunate, and that he should stay a few years more in Yedo to master astronomy and then return home to devote himself to observation. This proposal, however, was opposed by many and withdrawn without being put into effect. As they agreed to have Tadanori go home at once and come up to Yedo once in a while to study with Kageyasu Takahashi, he went to his home in Sawara at the end of 1822, taking with him Inō's field-notes and other books connected with surveying which he had in his charge. About this time an epitaph for Inō was composed by Issai Satō, a great scholar of Chinese classics and a professor at the Shōheikō (Academy) in Yedo, and this was carved on a stone which was erected over Inō's grave in Genkūji. The tombstone still stands there with the epitaph, which has, in spite of a few errors and the probable omission of some letters, served as a chief means of making Inō's career and achievements known to the public.

After returning to Sawara, Tadanori set up in his court several instruments for astronomical observation that had been handed down from his grandfather. He engaged himself in observing the position of the sun for certain periods, the solar and lunar eclipses, and other special phenomena. He also came to Yedo once or twice each year for two or three months to further his study of astronomy. He died in the spring of 1827, at the early age of twenty-one.

Inō originally had four sons and three daughters. Kagetaka, the eldest son (1766–1813), who had succeeded to Inō's position at Sawara when the latter retired, had died before his father, leaving two sons, Tadanori and Tetsunosuke. Takayoshi, his second son (1786–1838), who went by the name of Shūzō, became the son-in-law of Sakurai, but later returned and spent his ruined life in teaching elementary lessons to children, under the name of Genjirō Jimbō. Junji, the third son (1788–1794), and the fourth son whose name is unknown both died young. Ine, his eldest daughter (1763–1822) who married a man named Moriemon took the tonsure after her husband's death, assuming the Buddhist name of Myōkun, and returned home. Shino, his second daughter (1769–1788) married a certain Shūsuke Kase, but died soon after. Koto, his third daughter (1789–?) married one Jōemon Matsuda. Kagetaka was succeeded by Tadanori, his eldest son (1806–1827), who went by the name of Saburōemon. When Tadanori died, his daughter and Tetsunosuke, his younger brother (1810–1818) had already passed away before him. Thus the main family of the great surveyor died out. So one of the relatives was chosen as the successor to the family, but this gave rise to complications among the relatives until the property which Inō had replenished with great effort in his earlier years was endangered within a few years of his death. Fortunately, Tsuda, lord of the Sawara estate, regretting the fall of the noble family, ordered Kageharu Mozaemon Inō (Sekken by pen-name), one of the most influential relatives, to take charge of the whole property. This intervention alone saved it. Some thirty years after Tadanori's death, a relative was adopted to be the husband of a daughter of Mozaemon and they succeeded to Tadanori's property, and their descendants survive to-day.

6 The Effect of Inō's Survey-Maps of Japan

It goes without saying that the survey-maps of Japan which Inō had put his soul into, satisfied the demand of the official authorities and contributed much to the defence of the country. The original copy of the map of the whole country was stored in the library of the Shogunate and the copies were secretly valued by some feudal lords or others of special position. Since only a very limited number of people were allowed to see them, the maps were not made practical use of by common people. However, when in 1826 Kageyasu Takahashi, violating the national interdict, gave Siebold some copies of Inō's survey-maps in exchange for the books Siebold possessed and raised a great scandal, the name of Inō's survey-maps created a sensation afresh throughout the country and the maps Siebold secretly took abroad with him came to be valued as rare objects among certain scholars who admired the excellent technical skill of the Japanese astronomer. As soon as Takahashi's crime of giving the maps to Siebold was revealed, the Shogunate seized Siebold's luggage and believed that they had confiscated every document prohibited by the law. As a matter of fact, however, Siebold had copied the principal maps during the night, scenting the imminent danger, and had concealed them with other important books and finally succeeded in taking them out of the country, as is told in his diary published in his book, "Nippon." The chief documents Siebold secretly took with him were Inō's map of the whole of Japan (except Yezo) drawn on a scale of 1:864000, Inō's map of Yezo Island drawn on a scale of 1:432000, a sheet of Inō's large map for Shimonoseki and its vicinity, a small map of the Japanese frontiers compiled by Kageyasu Takahashi, maps of Yezo, Saghalien, and Eastern Tartary drawn after the travels of Rinzō Mamiya and Tokunai Mogami, and Rinzō Mamiya's "Travels in Eastern Tartary." Siebold, by the way, did not know at that time that the maps had been drawn by Inō; and he simply called them the Government maps of Japan.

On his return to Europe, Siebold sent Inō's map to Adam John Krusenstern, an eminent authority on hydrography at that time, asking for his criticism, in order to have it estimated more accurately. Krusenstern wrote an answer dated Oct. 23rd, 1834, in which he said:

".... Ohne mich in eine genaue Analyse derselben einzulassen, will ich hier nur bemerken, dass nach einer Vergleichung mit unseren europäischen Karten, d.h. mit solchen, welche nach einer genauen Untersuchung von bekannten Seefahrern konstruiert sind, ich nicht anders als sehr günstiges Urteil über sie fällen kann. Nicht nur die Konfiguration der Küsten und das Detail derselben, wo solches auf unsern Karten zu finden ist, sondern auch die Längen und Breiten stimmen auf eine wundervolle Weise und liefern einen interessanten Beweis von den Fortschritten, welche die Japaner in der Astronomie gemacht haben. . . . . Bei dieser ausserordentlichen Genauigkeit, welche sich fast überall zeigt, wo ein Vergleich mit europäischen Seefahrern zulässig ist, lässt sich wohl mit Gewissheit annehmen, dass auch diejenigen Küsten, welche bis jetzt nicht haben von europäischen Seefahrern untersucht werden können, den nämlichen Grad der Genauigkeit gewähren, und dass folglich die Lücken, deren es bis jetzt noch viele an jenen Küsten giebt, durch Ihre Karten genügend ausgefüllt werden können, bis endlich ein wissenschaftlicher Seefahrer dort das Werk vollendet, wozu dem Anscheine nach jetzt keine Aussicht ist. . . . . so wird man gewiss auch dort mit Ungeduld der Herausgabe Ihrer Karten entgegensehen. Es wäre daher sehr zu wünschen, dass dieses so bald als möglich geschehe. . . . ."

The existence of such an accurate survey-map in Japan astonished Europeans. It may be said that Inō's map demonstrated the latent power of the Japanese scientists. This map of all Japan, contracted and replotted on Mercator's projection, was published by Siebold in 1840 and was introduced to European scholars. The map which Siebold took home was a duplicate hastily copied from a map of a small scale and as this was printed on a still more reduced scale, which, at the best, merely indicated the correct position and shape of the Japanese islands. The fine indentations of the coast-line, showing evidences of Inō's pains, were not reproduced thereon.

In 1861, the British Navy undertook a survey of our coast, having procured permission from the Shogunate Government, though this was granted under compulsion. It was only when the captain of the British surveying vessel happened to see a small-scale map of all Japan that the true value of Inō's map came to be recognized by foreigners. The expulsion of foreigners was strongly agitated at this time, and such work as a coast-survey by foreigners was seriously feared since it might readily give rise to complications. Although the Shogunate under pressure permitted the English to survey the coast, the authorities were extremely concerned lest troubles should arise; they despatched instructions to the clans along the coast, sent some Shogunate officials aboard the British cruiser for protection and supervision as well as for negotiation with the officials of different clans. They took pains to prevent complications with local officials or people. About ten days after the British cruiser commenced a survey from the mouth of Yedo Bay, the captain chanced to have a look at Inō's small-scale map in the possession of the Shogunate officials on board and, comparing the result of the new measurements with this, he realized the remarkable accuracy of Inō's map. The Englishmen, being now convinced of the uselessness of re-surveying the coast-line, gave up their original plan to conduct a survey from Yedo Bay down the southern coast and round the coast of Kyūshū to Nagasaki, and only sounded the principal ports in Awa, Idzu, Sagami, Suruga, Tōtōmi, and Kii, and determined the position of some reefs and the longitudes and latitudes of some points on land. It may be said, therefore, that Inō's map served to prevent unavoidable complications between the local officials or people and the Englishmen and consequently saved the national prestige from injury. Not long after this, a chart of the Japanese coast was published in Great Britain, with the note that the coast-line had been based upon the original survey by the Shogunate Government. The map which the Englishmen obtained was a copy of the map on a scale of 1:432000, and it is still carefully preserved in the library of the British Admiralty.

In this way Inō's maps not only contributed to foreign relations in the last stage of the Shogunate, but also proved of much practical use in domestic affairs—such as in navigation and transportation; and especially as our navy was newly established and the art of navigation was being studied, these maps served as the sole guide for the coast navigation of our country. The small-scale map was printed at the Kaiseisho, the precursor of the Tōkyō Imperial University, at the time of the Restoration. Even in the Meiji era the authority of Inō's maps was never questioned, but more and more recognized; the map of Japan drawn at the Geographical Bureau of the Government to be exhibited at the Vienna Exposition in 1873, the map of Japan compiled by the Department of Education in 1877, and the map of Japan made by the Geographical Bureau of the Department of Home Affairs in 1880, were all based on Inō's maps, and especially the coast-line was plotted entirely according to them. Even the maps on a scale of 1:200000 compiled and published in 1886 and several succeeding years by the Military Land-Survey Department, which were to meet with the pressing demand of the public before the recent survey was completed, owed much to Inō's survey in the matter of coast-line.

Inō's survey-work gave considerable convenience not only to the Shogunate at the time it was accomplished, but also to the Japanese Government and people for a number of years, as well as to the navigators of the world. Sixty-five years after Inō's death, in 1883, the Emperor Meiji conferred upon him posthumously, for his important services, the Senior Grade of the Fourth Rank. His Majesty was also pleased to inspect the maps Inō drew and the measuring-instruments he employed. The Tōkyō Geographical Society erected in 1889 a bronze tablet at Maruyama in Shiba Park, Tōkyō, in order to celebrate Inō's stupendous acheivement. And the people of Sawara where the Inōs still live built in 1919 a bronze statue at Suwa Park there to commemorate this great man their town had produced.

CHAPTER IV

THE METHOD AND INSTRUMENTS EMPLOYED BY INŌ IN HIS SURVEYS

In fulfilling his purpose of surveying Japan, Inō did not adopt any new method unknown to the land-surveyors of his day. The principal points in his improvement upon the surveying of former times were that he made the utmost use of the method of sighting distant mountains, a practice of former land-surveyors, in his efforts to correct and adjust errors in surveying; that, in order to correct and adjust errors in surveying wide areas, he personally conducted astronomical observations to determine the longitudes and latitudes of a number of stations, which had always been considered a mere theory and never practised by Japanese land-surveyors; and that he greatly increased the accuracy of surveying by the improvement of measuring-instruments and by the utmost care in using them. These should be considered the chief factors that made it possible for Inō to carry out the survey of such wide areas as had defeated his predecessors, and to fulfil his purpose of constructing a complete map of Japan.

1 His Method of Land-Surveying

The method of land-surveying adopted by Inō was what is to-day called the method of traverse—one most commonly practised at that time. In applying this to practical use, however, he had measuring-instruments so precise as had never before been seen, and took the utmost care in order to increase the accuracy of his surveys. The actual processes may be roughly known from "Inō-Tōkō-Sensei-ryū-ryōchi-denshūroku" (The Record of Training in Land-Surveying of Master Tōkō Inō's School) compiled by Shin Keijirō Watanabe, his aptest pupil.

On employing the traverse method, this book says, "In surveying a highway, a coast-line, an island, or a lake, the first stake is driven into the ground at the starting point; and then at each bend of the highway or coast-line a stake is driven into the ground where a pole with a piece of paper on its top is planted; the distance between the stakes is measured with an iron chain or a rope made of bamboo or rattan; the azimuth from one stake to the next is doubly measured by fore-sighting and back-sighting with two small magnetic compasses; the man who has measured the distance writes down its numerical value on a small piece of paper and hands it to the coolie who holds the pole, and the man who has measured the azimuth writes down the value on the paper on which the distance has been registered; at that time the angles of fore-sighting and back-sighting are compared to test whether an inadmissible error or mistake may have crept in. The distance and azimuth on the piece of paper are entered in order by the recorder in the field-note. These pieces of paper are preservd in order on a bamboo skewer so that they may be checked with the field-note after surveying every day. In measuring the circumference of a lake or an island, the line of survey around it should be closed connecting the last station with the starting station. Where a projecting cape with cliffs and precipices admits no accurate survey, a careful survey should be made across the cape.

"The inclination of a hilly road should be measured with a quadrant 20 cm. in radius; and the distance measured along the slope should be multiplied by the cosine of the inclination and reduced to the horizontal distance. In surveying an island, a large pole with a piece of paper on its top should be planted at each of the capes of the island and other such poles at suitable points on the mainland, and the position of the island is determined by the method of intersection, measuring the mutual azimuths between the poles; when an island is distant, the summits of mountains may be used instead of the poles. While measuring a highway or coast-line, the azimuths of trees, buildings near by, hills and mountains far and near should be measured by small magnetic compasses from proper points so that their positions should be determined by the method of intersection. As this will be extremely valuable in adjusting errors in map-making, mountains which may be used as marks should be sighted from as many points as possible. The width of a river which cannot be surveyed directly should also be measured by the method of intersection."

This was exactly the method which he adopted in his surveys under ordinary conditions, and by which he expected to secure sufficient accuracy in surveying. Where the coast was too precipitous for passage and admitted no proper means of surveying across, a small boat was used in stretching a measuring-rope along projecting rocks, and where even this was impossible, the topography was sighted by the method of intersection from a distance. Such measures were taken according to circumstances.

Inō applied the method of intersection and the method of cross-cut survey more extensively; whenever he came to a point affording a great range of vision, he measured distant mountains of comparative prominence with magnetic theodolites of various sizes; and he made a survey of the highways running in many directions, which he connected with the coast-line. All this served him in correcting and adjusting errors in the survey of more extensive areas.

In this way Inō never plotted a map directly on a plane-table. The distances and azimuths measured with a measuring-rope and a magnetic compass were minutely written in numerals, together with the description of the topography of the route, in the field-note. Afterwards a map was drawn from this field-note. The field-notes were, therefore, Inō's only fundamental source.

It is a great misfortune that all these valuable original documents, except fragments of a few sheets, have been scattered or destroyed by fire.

The field-notes have been lost, except fragments of a few sheets, it is true, but there still remains a record of the azimuths of distant mountains and islands Inō observed from suitable points of the surveyed routes for determining the position of the mountains and islands and also for reference in adjusting errors; this, being an extracted fair copy of the field-notes, is entitled "Santō-hōiki" (The Record of the Azimuths of Mountains and Islands) and consists of 67 volumes representing nearly all parts of Japan. According to this record, Inō made observations carefully with more than two (commonly three or four) magnetic theodolites that the instrumental error might be reduced to a minimum. But he was satisfied with observing azimuth from one direction: that is, he never counter-measured the azimuths of the points on the surveyed route from the top of the mountain or island.

A mere description of the topography along the surveyed routes entered in the field-notes, Inō thought, was not quite sufficient in plotting the topography on a map. And when he surveyed the western regions of Japan, one of his party was engaged in sketching the general features of towns, villages, mountains, rivers, and seas along the route, so that his map might be true in representing the geographical features. A few fragments of such rough sketches made on long sheets of paper are still preserved to-day.

2 His Land-Surveying Instruments: Their Construction and Use

It is evident that Inō possessed some surveying instruments while in his native place. But the instruments he used in surveying the country were all constructed after he came up to Yedo, under Yoshitoki Takahashi's instructions and after a profound study of various instruments. The more important among these were the measuring-scale, the measuring-pole, the measuring-rope, the perambulator, the magnetic compass, the magnetic theodolites of three sizes, large, middle, and small, the theodolite of a semi-circular form, and the small quadrant as an inclinometer. Some of these are still preserved; the others have been lost, but their construction may be inferred from documents.

(*A*) *The Measuring-Scale*

The instrument which constitutes the basis of surveying is the measuring-scale. The Japanese system of scale has undergone frequent changes since remote times. The ***shaku***, the unit now generally used, originated from the Daihō Statute issued in 702 A.D. However, the standard measure as it was handed down generation after generation came to be somewhat diversified. In the early years of the 18th century the Shogun Yoshimune Tokugawa undertook to standardize it again. Consulting ancient documents extensively and considering its origin, he constructed a standard scale, a duplicate of the so-called small scale of Daihō, preserved in the sacred store-house of the Kumano Shrine in Kii Province. This was called "Kyōho-shaku." But as the Shogunate permitted the people to manufacture measuring-scales freely and imposed no restriction whatever, that used among the people was not yet unified. There were two kinds most widely used. One was the "Nembutsu-shaku" manufactured in Kyōto, and its length was equal to the Kyōho-shaku. The other was the "Matashirō-shaku" which was commonly used among carpenters, and its length was shorter than the Kyōho-shaku by 4/1000.

According to a tradition which was widely circulated and which appeared several times in literature, the scale Inō used in his surveying was made by compromising the Kyōho-shaku and the Matashirō: that is, it was shorter than the former by 2 *rin* and similarly longer than the latter, and it was, therefore, called the "Setchū-shaku" (the compromised scale). In the manuscript of a book compiled by the Department of Agriculture and Commerce, it is said that, when weights and measures were fixed by law in this country in 1875, Inō's scale was adopted as the legal standard, and that its length corresponded to 0·30303 m. when it was compared with the metric scale in 1880. If we accept this statement, the length of Inō's scale exactly coincides with the length of our present legal unit.

What was the temperature at which the length of the Setchū-shaku was taken as the standard? Where was the Setchū-shaku found? As to these important questions, no authentic record gives an answer. The fact that the Department of Agriculture and Commerce possesses three scales, or the alleged imitations of the Kyōho-shaku, the Matashirō-shaku, and the Setchū-shaku, which had been in the possession of Gokwan Uchida, a mathematician of the last days of the Shogunate, together with a brief history of the scales written by him, no doubt proves that Uchida was responsible for the opinion that Inō made the Setchū-shaku by compromising the two scales and that the length of the Setchū-shaku the Government adopted was presented by Uchida.

The scales which were in the possession of Uchida are all of brass and 2 cm. wide; as to their thickness, it is about 2 mm. at one edge, but from the centre it gradually decreases towards the other edge where it is less than 1 mm. The graduation is marked at the thinner edge. Both ends of the scale terminate by lines, so that the scale has the form of a line-measure; the interval between the two ends is graduated into hundredth parts with parallel lines, and the ten parts at either end are subdivided into five equal parts.

In the autumn of 1910, or before these scales were destroyed in the earthquake and fire, they were compared with the metric scale and the result was as follows:—

	Length at 15°C.	Length at 0°C.
The Kyōho-shaku	0·30365 m.	0·30357 m.
The Carpenter's shaku	0·30237 m.	0·30229 m.
The Setchū-shaku	0·30305 m.	0·30297 m.

N.B. The coefficient of linear expansion of the scales was each supposed to be 0·0000183.

According to this result also, the length of the Setchū-shaku was about an average of the Kyōho-shaku and the Carpenter's, almost equal to the *shaku* of the present system at ordinary temperature. If this Setchū-shaku is an accurate imitation of the scale Inō used, the value mentioned here and the comparison table given in a document in the possession of the Department of Agriculture and Commerce, we must admit, doubly show us the length of the scale which Inō adopted in his survey of Japan.

Was the Setchū-shaku really designed by Inō in surveying as a compromise of the Kyōho-shaku and the Carpenter's? And did the imitation scale formerly possessed by Uchida and now by the Department of Agriculture and Commerce really correspond with the system that Inō adopted? We have our doubts on these questions.

In the preface to Inō's small-scale map drawn in 1804, Kageyasu Takahashi writes, "For further reference in future, I submit to the Shogunate with the maps the magnetic compass which Inō designed and the scale which Inō used in drawing these maps." From this it is evident that in his surveys Inō exclusively adopted the scale of a particular description as he designed an accurate magnetic compass. As to the magnetic compass, Takahashi describes in this preface that Inō constructed such an accurate instrument by his own design, but it is a matter of great regret that he does not say a word about the history of the scale. And the scale submitted to the Shogunate has been lost to sight. In the preface he mentions that the length of one French *toise* corresponds to our 6·42 *shaku*. If this comparison is accurate, Inō's scale may be converted to the present system: the length of Inō's scale is thus computed to be 0·30359 m. according to the metric system, or 1·0018 *shaku* of the present system. The comparison of the *toise* and the *shaku* conducted by Takahashi could not have been very accurate, we are afraid, and, therefore, this result may not be relied upon. However, even if we allow an error of ±0·005 *shaku* in the comparison, Inō's scale will be 0·30335–0·30382 m. by the metric system, or 1·0011–1·0026 *shaku,* which as is seen, differs at least by more than 0·001 *shaku* from the result obtained for the Setchū-shaku.

Moreover, the unification of scales is not only important from the economic standpoint of the nation, but is also a matter of supreme importance in the field of science as it develops. Inō's surveying was not the only thing that demanded the solution of this problem. One of the reasons why the Shogun Yoshimune Tokugawa established the Kyōho-shaku after a study of ancient scales must have been that he keenly realized the necessity of defining the scale in graduating the gnomon he himself constructed. As the study of astronomy developed gradually in Japan towards the end of the 18th century, the problem of defining the scale had to be solved from the scientific standpoint. Shigetomi Hazama, for instance, made a study of the history of the scales in our country and the comparative lengths of the Japanese and foreign scales. The scale adopted for graduating the gnomon at the Astronomical Bureau which was constructed chiefly under the supervision of Yoshitoki Takahashi and Shigetomi Hazama at the time of the compilation of the Kwansei Calendar—the scale should have been based upon some good authority and decided upon after much careful consideration, and all the scales employed at the Astronomical Bureau must have followed this scale system. If so, it is very improbable that Inō, a scientist instructed and assisted by Takahashi and Hazama, should have freely modified in his surveying the scale adopted at the Astronomical Bureau of the Shogunate. In my opinion, the Setchū-shaku of Inō as we call it to-day was the scale Takahashi and Hazama first adopted in constructing instruments at the Astronomical Bureau, and had little connection with Inō though he was misrepresented as the originator of the new scale by some people when he used it widely in his surveying. May this not be how it came to be known as Inō's Setchū-shaku?

After all, Gokwan Uchida's statement that Inō constructed the Setchū-shaku is hardly to be relied upon, and the alleged copy of the Setchū-shaku formerly possessed by Uchida and lately preserved by the Department of Agriculture and Commerce may not be accepted certainly as representing the scale Inō adopted unless its history be made clear.

Is it, then, impossible to determine the length of Inō's scale to-day? Fortunately we have discovered, in recent years, two scales in the godown of the Inō family that must be considered as Inō's original measures. They are both made of brass and are constructed almost alike. One is a brass plate 42 mm. wide, 4 mm. thick, and 360 mm. long and is fringed by a brass bar of 5 mm. × 4 mm. in section screwed down along one edge; on the surface of the plate, eleven lines are drawn at an interval of 3 mm., parallel to the length of the plate; and at right angles to these parallel lines, parallel lines are drawn from one end of the plate at an interval of one *bu* to the length of one *shaku*; and then diagonals are drawn between these successive graduations, and thus the length is subdivided into *rin* or thousandth parts of a *shaku* by the intersection of the parallels and diagonals. The scale is provided with a slide index which may slide freely on the scale so that one may estimate one tenth of one *rin* by sighting the intersection of one edge of the slide and diagonals. This scale has a particular form of graduation in that one end is limited by an end of the instrument while the other end is limited by a line. On the back of the measure is engraved Tanryū Tanaka, the name of the maker.

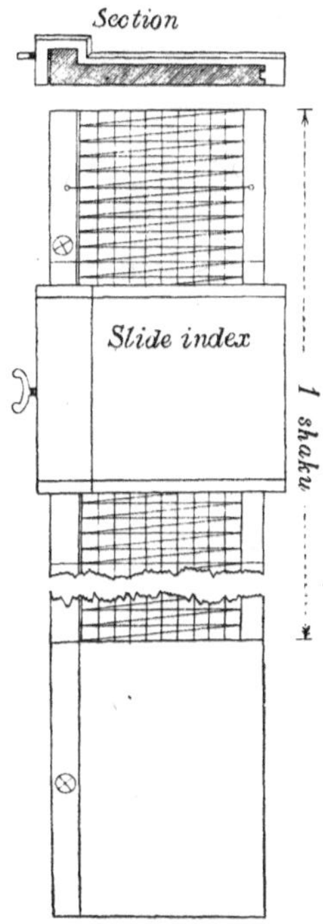

Fig. 1.

The other one is made of a single brass plate 38 mm. wide, 3·5 mm. thick, and 345 mm. long; on the surface, like the other one, eleven parallel lines are drawn at an interval of 3·5 mm.; lines are drawn across them at right angles at an interval of one *bu* to the length of one *shaku*, the diagonals are also drawn and it is provided too with a slide index. This differs from the other only in that this is a perfect line-measure limited by lines at both ends. The graduation is much more clearly engraved, but the name of the mechanician is not engraved.

Not only from their appearance, but also from the rich brocade bag in which they were carefully preserved, it is evident that these are the scales which were immensely valued by Inō and are what should be called his original measures. It is probable that previous to the commencement of his survey-work, Inō had these scales made as part of the gnomon when he had a number of astronomical instruments constructed through the intercession of Shigetomi Hazama and others. This may be proved by the fact that these scales exactly correspond in form with the scale accompanying the gnomon described in the "Kwansei-rekisho-gishōshi" (The Description of Astronomical Instruments Used for the Compilation of the Kwansei Calendar in 1797), and that Inō's letter to Hazama concerning the price of the appendage to the gnomon still exists.

Which of these two scales Inō regarded the more important, we are unable to decide. It would seem that no particular difference between the two scales was observed at that time. We should acknowledge them as of the same value. The writer standardized them in July, 1909, and the following result was obtained:—

	Length at 15°C.	
The Scale by Tanryū Tanaka	0·30325 m.	1·00073 shaku
The Anonymous Scale	0·30336 m.	1·00109 shaku

The difference of length between the two scales is 0·00036 *shaku* or 1/3000 of the whole length. Though too great an error to be overlooked to-day, such a difference was perhaps taken for granted more than a hundred years ago when the construction of instruments depended not so much upon mechanism as the skill of mechanicians. Though these scales are so constructed as to enable one to measure with the naked eye to a ten-thousandth of a *shaku* by means of the slide index, the point of intersection of the edge of the slide index and a diagonal is not certain as they intersect with a small angle. On this account, when these measures are used, it seems quite natural that an error of two or three *mō*, or ten-thousandths of a *shaku*, is introduced, so that such an error of 1/3000 of the whole length was always ignored as unavoidable.

What confronts us next is the question—What was the temperature at which Inō adopted these original scales as the standard length? The fact that the volume or length of matter changes with the temperature was thoroughly known to the Japanese scholars even before 1800. This problem is discussed in the letters between Takahashi and Hazama. Thermometers had been imported from an early date and about 1800 there existed some which had been constructed in Japan. From Inō's diary, it is evident that either he possessed one or had the privilege of keeping one at hand.

Since the expansion or contraction of matter with the change of temperature was known, and an instrument to measure temperature accessible at that time, naturally the temperature of a scale ought to have been discussed in determining a standard length. As we peruse the "Ryōchi-denshūroku" and other documents, the influence of moisture upon the length of the measuring-pole and the measuring-rope is carefully studied and precautions against this are given in various places, but nothing is said about the influence of temperature or about precautions against it. It seems that Inō did not consider the temperature effect throughout his life. It may be due to the fact that the coefficient of the linear expansion of ordinary metals, wood, or bamboo was trifling and the influence of the natural temperature of each season upon the expansion or contraction of such materials was not great enough for Inō to consider the influence of temperature upon the scales to which he had already allowed about 1/3000 *shaku* for an experimental error.

As Inō's scales were all of brass, the coefficient of their linear expansion was approximately 0·000018; and if we allow 30° for the difference of temperature in summer and winter, one *shaku* scale of brass will cause a difference of 0·00054 *shaku* with this change, or an expansion or contraction of 0·00027 *shaku* from the length at an average temperature. Such expansion or contraction was completely ignored. Thus, Inō ignored an error of 1/3000 of the whole length as to the absolute length of a scale or the effect of temperature. Hence as our conclusion on the standard scale Inō adopted, basing our argument upon the two scales preserved by the Inō family, we are in a position to assert unhesitatingly that the *shaku* that Inō adopted was at ordinary temperature approximately 0·3033 m., or 1·001 *shaku* of the present system; and that, according to the difference in scales or in temperature, the scale was longer or shorter by approximately 1/3000 *shaku* or less.

(*B*) *The Measuring-Pole*

The measuring-pole which Inō used does not exist to-day. It is impossible to know of its details. But the documents extant represent it as one *ken* (that is 6 *shaku*) in length and so constructed that two poles might be connected to obtain a length of two *ken*.

As to its accurate construction, no document offers any data. To borrow analogy from other measuring-instruments, this instrument was no doubt made of the fine grain of an old Japanese cypress that it might not warp, and both ends were capped with brass. It seems that Inō used the measuring-pole not so much as a direct instrument for surveying, as one for testing the measuring-rope, etc.

(*C*) *The Measuring-Rope*

The measuring-rope is the most important instrument for surveying distance in field-work, and many land-surveyors before Inō had studied the construction of this instrument. A measuring-rope made of ordinary material was by no means satisfactory, because of its considerable change of length in moistness or dryness. When he commenced his survey-work, Inō spared no pains to improve this. At length he found it best to use an iron chain, and this he used generally; however, Inō had many ropes made of different materials, and he adopted the most convenient that each occasion demanded.

The iron chain was made of iron wire cut to the length of approximately 45 cm. with both ends bent into rings so that their inside measure was 1 *shaku*; 60 pieces were connected to make up an iron chain for 10 *ken*. Though its length was not affected by the moisture, the chain, when drawn over for a long distance, was likely to make some slight curvature at each segment. Its length was also to be affected, not only by the wearing down of the links with which segments were joined together, but also by the gradual deformation of the links into ellipses. When he used the iron chain, therefore, he made it a rule to examine its length quite frequently. The rattan-rope was made of a few rattans imported perhaps from Formosa and pieced together to a convenient length. The whale-rope was made of several dozen of whale fins split to the width of about 2 cm. and pieced together. The bamboo-rope was made of bamboo cane of from 15 to 20 cm. in circumference cut to the length of 4 m. and split into eight strips and then pieced together. The length of the rattan-rope, the whale-rope, or the bamboo-rope was 5 *ken*, 10 *ken*, or 20 *ken*. For their lighter weight and greater handiness than the iron chain, these were much favoured as auxiliary measures.

Sometimes a rope made of hemp fibres varnished with *shibu* or the juice of unripe persimmons was used. Whenever used, its length was examined before and after surveying. During the first year of his surveying, Inō was unable to use a measuring-rope because of his attempt to survey a long distance in a short period and because of the lack of the official co-operation that he desired, and distance was only roughly surveyed by pacing.

(*D*) *The Perambulator*

The perambulator is an instrument by which the distance covered is computed by the number of the revolutions of its wheels rolled upon a surveyed line. This was used from the earliest days in China and was for a long time supposed to be a handy surveying-instrument. When Inō started his survey-work in 1800 he possessed a simple wooden perambulator, but he did not use it. The one with brass wheels constructed under the direction of Takahashi in the following year, Inō used as an auxiliary surveying-instrument in measuring some highways.

Fortunately this perambulator is preserved by the Inō family. As compared with ones described in ordinary books on land-surveying, it is far more delicate and complete in its construction. As shown in the figure, the body is a stout box 24 cm. wide, 21 cm. deep, and 34 cm. long, made of hard wood; in the hinder part there is a horizontal iron axis to which two brass wheels are fixed and the axis revolves with the wheels. The wheels are each about 20 cm. in diameter and 2 cm. in thickness, and stand parallel 12 cm. apart. To the fore part of the box are attached two small wheels, 6 cm. in diameter and 1 cm. in thickness. These four wheels, standing parallel, are so fixed that their course may not be freely changed; when drawn forwards by a rope fastened to the body, it always moves on in a straight line as long as the wheels do not slide sideways.

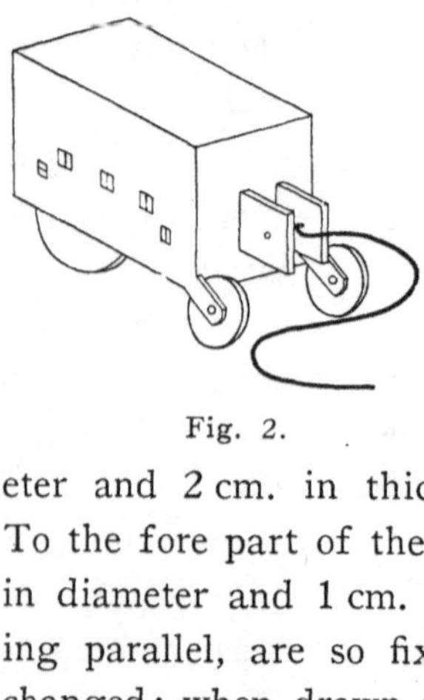

Fig. 2.

There is a contrivance to remove mud, dirt, etc. from the larger wheels. As shown in Fig. 3, several index wheels connected with one another by toothed wheels are so arranged that they revolve one after another with the revolution of the axis of the larger wheels, and the distance covered by the larger wheels is readily read by means of the figures and indexes on the index wheels. The instrument is so constructed that one revolution of the larger wheels corresponds to a distance of approximately two *shaku*; on the first index wheel are graduated the figures from 0 *ken* up to 10 *ken*; on the second index wheel tens; and so on till on the fifth index wheel are marked ten-thousands; and from 100000 *ken* the graduation goes back to 0 *ken* again.

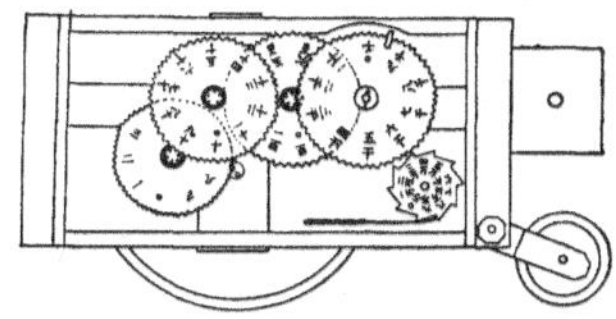

Fig. 3.

The perambulator is a very handy measuring-instrument which when drawn upon the ground tells the distance it travels or what is proportional to this by the graduation on the index wheels, provided that the surface of the earth is even, and the larger wheels do not slip, or if the amount of slipping remains constant. When put into practice, however, the revolution of the wheels varies with the hardness, the moistness as well as the roughness of the ground; not only do the figures on the index wheels fail to indicate the real distance surveyed, but also the relation between the figures indicated and the real distance always varies, depending upon circumstances, and thus is never constant. The perambulator seems to have been first constructed to be used in surveying not only highways, but the coast-line also. As it was completed, Inō tentatively surveyed part of the Tōkai Highway with it in the summer of 1801 and also tested it in surveying the coast-line of the Bōsō district. Within a few days, Inō realized the gross inaccuracy of the figures indicated by the perambulator and tried various means of rectification. But the smooth seashore was generally so sandy, and hard ground so rough that Inō learned that the perambulator was of no use in surveying the coast. For this reason it was used only as an independent measuring-instrument, during 1802 and 1803, in surveying short distances within castle-towns of feudal lords or highways branching off from the main surveyed line to castle-towns; it was necessary to show some respect to the feudal lords and yet this deference did not have any appreciable influence upon the accuracy of the whole survey. In other cases, the perambulator was always used simultaneously with the iron chain or rattan-rope only as a counter for confirming how many times the measuring-rope was connected from one station to the next one. In the survey of western Japan which was carried on after 1805, its use was entirely abandoned.

(*E*) *The Magnetic Compass*

Though the magnetic compass had been used in Japanese land-surveying for a long time, the one used before Inō's time generally had a short flat magnetic needle and was ornamental in shape; and its magnetic axis seldom coincided with the axis of its geometrical shape. Besides, the cap was a piece of brass with a deep hole, which was hung on a dull-pointed pivot; the friction on the point of contact was really considerable. Therefore, there was always some difference between the direction a compass indicated and the direction of the magnetic meridian. In measuring a certain direction at a given point, the result obtained by different magnetic compasses usually differed by several degrees, and frequently the same instrument showed a great difference at different times. But most surveyors only took this to be the common fault of magnetic compasses. Some surveyors even valued a compass which settled down quickly on account of the great friction at the pivot, thinking it a convenient one in practical surveying. Among other surveyors, there were some who, considering the cause of the erroneous indication of a magnetic compass, advocated that, to assure greater accuracy in its indication, the compass should have a more slender needle and be made lightly moving. Inō realized this principle in his earlier years, as seen by the magnetic compass installed in a wooden theodolite which is supposed to have been used by him while in Sawara. This magnetic compass has a slender magnetic needle 4·5 cm. in length, and remarkably improved, as compared with the ordinary magnetic compasses widely used at that time.

Certain books imported from China in the Ming and Ching dynasties had mentioned the fact that the magnetic compass did not indicate true meridian; and not a few Japanese astronomical books, written after this, such as the "Hōreki-rekisho," also offered the same proposition. A magnetic declination of more than 4° at Peking was described in these foreign works; but the experiments made in Japan did not give such great magnetic declination as a whole, although various compasses gave some slight difference. Whether the magnetic declination mentioned in the foreign books proved the existence of such a phenomenon or merely the crudeness of magnetic compasses was now questioned. No one thoroughly studied this question, however, with the result that it remained only doubtful for more than half a century until the end of the 18th century. When Inō came up to Yedo to study with Yoshitoki Takahashi, he considered the improvement of the magnetic compass an important factor in the method of land-surveying and as the only key to solving this half-century-old problem. He did his best in constructing a suitable magnetic compass. If a magnetic needle was shaped like a long needle to serve its purpose better, it would be difficult to magnetize the needle strongly and consequently even a slight friction at the axis would readily affect the direction of the needle; and it seems that Inō's great painstaking was in minimizing such friction. Shigetomi Hazama, in the spring of 1798, saw in Ōsaka a Dutch celestial sphere with a magnetic compass fitted in its frame, the needle of which had a cap made of quartz instead of brass as was customarily used in Japanese needles. Hazama drew a diagram of the cap and gave it to Takahashi. Thereupon, Takahashi and Inō asked Hazama to have a quartz cap made according to this plan by a mechanician in Ōsaka; Inō fitted it in a magnetic needle in the form of a long slender needle; and he studied the comparative merit of differently shaped cavities made in the quartz and of the sharpness and bluntness of the pivot. As a result of this, he discovered the advantage of having a hard steel needle for a pivot, and of keeping its end as sharp as possible. He then produced a megnetic compass almost up to his ideal, which he used from the beginning of his survey of the country. The magnetic needles produced through the above-mentioned processes and employed by Inō have been preserved, fitted in theodolites, or kept at the Inō home as reserves and removed parts, and they number about twenty. These magnetic needles are all in needle form as shown in the diagram; and from 94 mm. to 102 mm. in length; and some are 0·5 mm. wide and 0·6 mm. thick at the centre, gradually diminishing in width and thickness towards the ends; and the others are 0·4 mm. wide and 0·8 mm. thick along the whole length. Though both kinds appear at a glance to be symmetrical, the south-seeking arm is made somewhat thicker than the north-seeking arm so that the dropping of the latter caused by the inclination of the magnetic lines of force may be balanced and the needle may always keep a horizontal position; and the south-seeking arm is painted red in order to be distinguished from the other. The cap is a hemispherical quartz 3·5 mm. in diameter with a

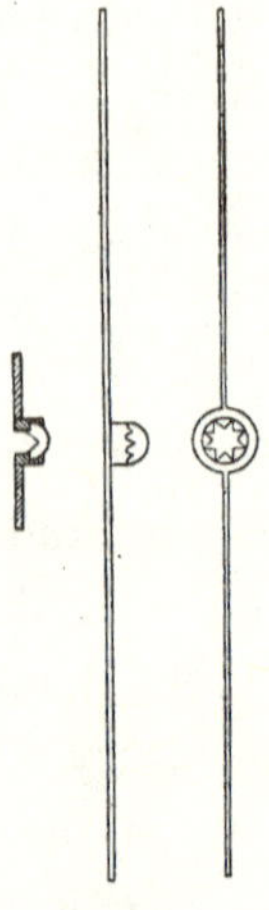

Fig. 4.

conic cavity having an angle of about 90° at the vertex; and it is fixed to the magnetic needle at the centre by the several catches of the brass washer inserted in the needle.

Inō also took every care in the manner of using a magnetic compass in order to realize its full effect. In his "Ryōchi-denshū-roku" he gives the following cautions. "When surveying, one should always carry a small whetstone, a file, a pair of pincers, and a few spare sharp pivots, to be used in sharpening the point of the pivot of the magnetic compass whenever it grows dull and the magnetic needle settles down too rapidly. When sharpened, the point should be stuck into a piece of hard paper; if thoroughly sharpened, it makes no sound at all; if no whetting makes any good, one should obtain a new pivot. When using a magnetic compass, one should by no means carry even the smallest piece of iron; it would certainly affect the magnetic compass and cause an error. One should then carry copper or bamboo swords."

Inō affirmed that, if the delicate magnetic compasses he had succeeded in producing as mentioned above were employed with strictest care, they all pointed one definite direction at a given point and there was no appreciable instrumental error. In order to solve the half-century-old problem, he immediately set about experimenting with these accurate magnetic compasses whether there was such a phenomenon as magnetic declination. Fortunately or unfortunately this was a time when the astronomical meridian chanced almost to coincide with the magnetic meridian in Japan, and it was impossible to detect the magnetic declination definitely even with Inō's accurate magnetic compasses. Naturally he made a great mistake concerning the existence of magnetic declination. From his own experience Inō readily asserted that that ordinary magnetic compasses did not indicate true meridian was due to the crudeness of the construction or else to carelessness in handling the instrument; he was convinced that accurate magnetic compasses handled carefully always indicated true meridian in our country; and consequently he came to doubt the existence of magnetic declination in general. However, he did not absolutely deny its presence in this country; he, of course, did detect some magnetic declination, such as is observed in a region abounding in volcanic rocks, but this he attributed to a local abnormality caused only by the iron contained in the rocks in the neighbourhood; he did not believe in the existence of magnetic declination in general for a wide range. Therefore, in his survey of Japan, Inō always determined azimuth with the magnetic compass; he made no distinction whatever between magnetic azimuth and astronomical azimuth, as he considered one identical with the other.

The magnetic declination in Japan being so extremely small during Inō's survey, his misunderstanding on this point left no appreciable bad effect upon his survey or map-making. His misunderstanding, however, cannot be justified. The azimuths entered as true azimuths in his field-notes and "Santō-hōi-ki" (The Record of the Azimuths of Mountains and Islands) were all magnetic azimuths, and more or less different from the true azimuths, as a matter of fact. If we determine anew the true azimuths of the mountains and islands of which Inō measured magnetic azimuths and compare the result with Inō's, his misunderstanding would be utilized in determining how small the magnetic declination in Japan was in Inō's days and how it was distributed in the country. Hence the "Santō-hōi-ki" in about 70 volumes still extant is a fund of valuable material in this bearing as well as a record of Inō's surveying and map-construction. The only great difficulty lies in how to identify the position of the stations at which Inō sighted the azimuths of the mountains and islands; even with the abundant material given in the whole book, it would involve tremendous difficulty to locate definitely only a few dozen proper field-stations scattered throughout the country.

(*F*) *The Small Magnetic Theodolite*

As he succeeded in improving the magnetic compasses, Inō, co-operating with Yoshitoki Takahashi, had various theodolites constructed. Of these, the small magnetic theodolite, also known as the magnetic compass mounted on the top of a stick, or a magnetic compass hung by a universal joint, is a very simple instrument. One of the reasons for his success in surveying Japan so speedily may be said to have been his invention of this handy yet accurate instrument. The characteristic of this small instrument consists in hanging a magnetic compass equipped with a pair of alidades upon a stick by means of a universal joint. The idea of fitting a magnetic compass on a stick was not original with Inō. As early as the beginning of the 18th century, a stick with the magnetic compass was in use; but it was only a crude one equipped with a magnetic-compass case on the top of a wooden stick; in order to use it, one had to plant it in the ground, and secure its verticality by the plumb-line attached to it so that the face of the magnetic compass might keep horizontal. It was by no means a handy instrument. Inō's small theodolite was no doubt suggested by this stick with a magnetic compass, though he applied to it the principle of the universal joint which had been used in some nautical instruments, in order to remove its chief inconvenience; and the magnetic compass was so arranged as to keep itself horizontal all the time, in spite of the inclination of the stick.

All the small magnetic theodolites Inō used were, except one or two, compasses with circles graduated in reverse order. A number of them were constructed during the period he surveyed the country; and several differing in size are still preserved. Of all the extant theodolites, the most authentic is the one which was first presented by Inō as a keepsake to his bosom friend, Seiyen Kuboki, and is now in the possession of the Department of Science of the Tōkyō Imperial University.

This theodolite with a circle graduated in reverse order has a main body, as shown in the diagram—a circular brass case 109 mm. in outer diameter, 18 mm. deep at the margin and 27 mm. deep at the centre. The bottom of the case is filled with lead in order to increase its stability, and the lead is covered with a silvered brass plate *AB*, which forms the base of the graduated circle. 6 mm. above the base is found a graduated circle *EF* 95 mm. in inner diameter. The circle is also of silvered brass, and on its face it is graduated to each degree simply dividing the circumference into 360 equal parts; this graduation is so clear-cut that the reading may be readily taken to 10′ by the naked eye. The magnetic needle is hung on the end of a steel needle *CD* erected at *C*, the centre of the base, and keeps the same level with the graduated circle. Its form is that of a long slender needle with a quartz cap, as stated under the item of the magnetic compass, and 94 mm. in length. As there is a space of only 0.5 mm. between the ends of the compass-needle and the margin of the graduated circle, there is little fear of making an error due to parallax in taking the reading indicated by the point of the compass-needle. A glass plate *G* is placed on the graduated circle as a cover, which is, in

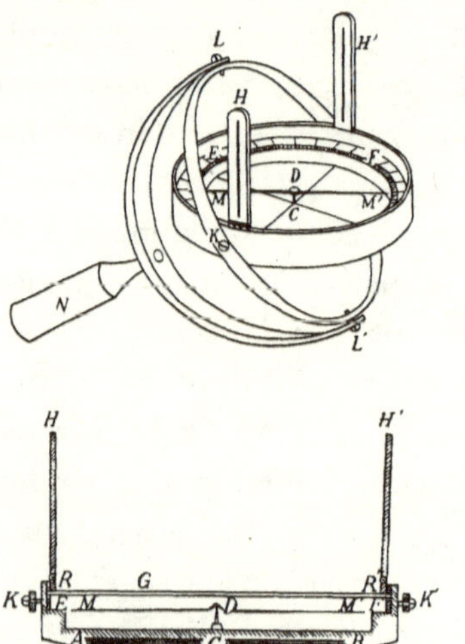

Fig. 5.

turn, pressed and fixed by a brass ring RR' 3 mm. thick and about 1·5 mm. wide in section. The brass ring at the same time forms the base of the alidades H and H', the bottoms of which are fixed so as to face each other in the direction of the meridian marked on the graduated circle. The alidades, the fore and the hind, are both exactly alike in shape and made of a brass plate 45 mm. long, 6 mm. wide, and about 1 mm. thick, erected vertically, in which a vertical slit 40 mm. long and about 0·3 mm. wide is made. The alidades are connected with their base by hinges that they may be folded into the case when not in use. The figures of the azimuthal degrees are marked, not directly upon the graduated circle, but on the inside of the glass plate, with Indian ink in reverse order. The universal joint on which the main body of the theodolite is suspended consists of a ring and a half-ring of brass. The ring is about 112 mm. in diameter, 6 mm. wide and 1 mm. thick in section; two brass screws facing each other connect the ring and the theodolite at K and K', at the side of the theodolite and immediately below the alidades, so that the theodolite may revolve about the axis KK' free from the ring. The half-ring is somewhat larger in diameter than the ring, just enough to cover the outside of the latter; in its section, it is also a little larger than that of the ring, and constructed very solidly; the half-ring is connected with the ring by two brass screws at two points, L and L', 90° apart from K and K' where the ring is connected with the theodolite; and this enables the ring to revolve free from the half-ring about the axis LL'. Thus, LL' and KK' constitute the universal joint: whatever position the half-ring may take, the theodolite, because of the weight of the lead in its bottom, will maintain its stability, naturally keeping the graduated circle horizontal, and consequently the slits in the alidades vertical. At the middle of the half-ring is fixed a brass handle N, which is cylindric, 21 mm. in outer diameter, 18 mm. in inner diameter, and 90 mm. in length. When used, this handle is mounted upon the top of the stick where it serves as an axis in rotating the whole instrument until the alidades take the desired direction. This theodolite weighs only 370 gm., including the universal joint. It is so small that it may be stowed away in a box 15 cm. wide, 24 cm. long, and 3 cm. thick —the most convenient of all the portable measuring-instruments.

The stick belonging to this instrument is a plain rod of oak 3 cm. in diameter and about 115 cm. in length; the upper end is made about 18 mm. in diameter for about 6 cm., so that it may be thrust into the hollow of the brass handle of the theodolite; the lower end has a metal ferrule with which the stick may be easily planted in the ground. However, a simple tripod was generally used to erect the stick upon; evidently it was only at the stations too steep for the tripod to be set up that the oak rod was directly planted in the ground.

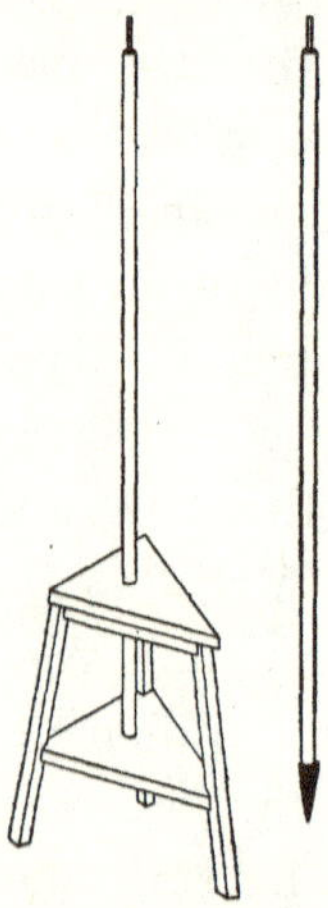

Fig. 6.

In measuring the azimuth of an object, the instrument is rotated on the top of the stick until the given sign-pole or mountain is sighted through both the slits of the two alidades, then the numerical figure of the graduation pointed by the end of the magnetic needle at such a position gives immediately the magnetic azimuth of the object. This is due to the fact that the graduated circle is fixed to the alidades that the two may rotate together, and the figures of the graduation marked on the circle is in reverse order. This is the reason why the instrument has been called a *gyakuban* or *gyakumeban* (reversely graduated circle). In azimuth measurement, Inō took two readings indicated by both ends of the magnetic needle to one-sixth of a degree by eye-estimation, and adopted their mean.

As stated before, the theodolites Inō used were mostly graduated in reverse order, but he also used the *jummeban* or the theodolite graduated in direct order. One of this kind still preserved shows its construction to be almost similar to that of the small theodolite graduated in reverse order; the only difference being that the graduated circle attached to the instrument is so constructed as to rotate freely and the graduation is in direct order; and a fine thread is strung in the direction of the graduated meridian.

To use this, the instrument is rotated on the top of the stick, as in the case of the other, till the alidades point to the given object, which is sighted through the slits, and the graduated circle only is turned until the fine thread strung on it coincides with the magnetic needle. Then two readings are taken by the indexes attached to the alidades immediately under the slits, and the mean of these readings is adopted as the azimuth of the given object. Thus, the directly graduated theodolite is much more inconvenient in handling than the reversely graduated one, nor does it by any means increase accuracy in surveying; evidently for this reason, Inō did not use the former except in special cases.

Though chiefly used in measuring the azimuth of the route surveyed by the method of traverse, these small theodolites were made use of in determining the azimuths of mountains close by, trees, temples, pagodas, and other special objects from various points of the surveyed route; and they were also used, together with the large and the middle-sized theodolites, in measuring the azimuths of distant mountains, islands, or any object whatever within the range of the naked-eye sighting.

This small theodolite suspended by the universal joint was used by Inō from the very beginning of his surveying in 1800. Because its handiness was proved by his practical work during the journey of that year, he constructed a few more accurate instruments the next year, and still more in the following years, which were used by his surveying party. He also made a good many at the request of his pupils. Theodolites built in the same form as the one Inō used and still preserved to-day number a dozen or more.

(*G*) *The Semi-Circular Magnetic Theodolite*

This was devised for decreasing the weight of a theodolite and simplifying its use. Two semi-circular magnetic theodolites are still preserved by the Inō family. As they are both almost alike in construction, we shall now describe one of them. This instrument consists of three main parts: a semi-circular brass plate which is graduated, a case in which a magnetic compass is fitted, and a pair of alidades. Above all, the graduated semi-circle forms the principal body, consisting of a semi-circular brass plate 170 mm. in radius, 4 mm. in thickness; at its centre a circular cavity *G*, 106 mm. in diameter, is made to receive the case of the magnetic compass; on either side of this, two cavities *I* and *J* are bored with a view to reducing the weight of the instrument. The graduation on this semi-circle is rendered on the so-called diagonal scale: with radii of from 166 mm. to 136 mm. seven concentric arcs $S_0, S_1, S_2, \ldots S_6$ are drawn along the margin of the semi-circle with C as their

Fig. 7.

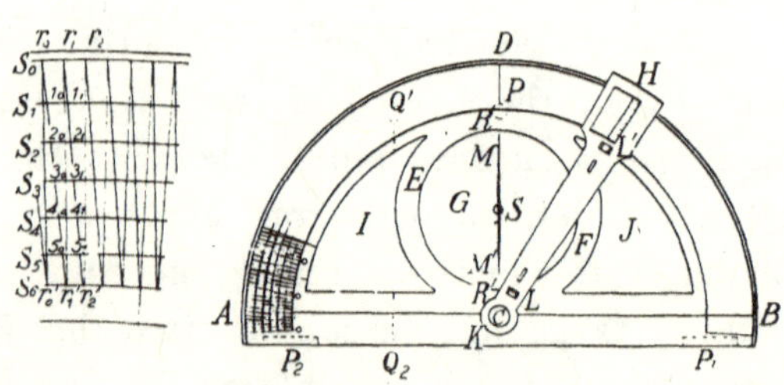

Fig. 8.

centre; these concentric arcs are graduated to each degree by the radii $r_0 r_0'$, $r_1 r_1'$, $r_2 r_2'$, etc.; and as for the interval of each degree, from the outermost arc to the innermost are drawn diagonals $r_0 r_1'$, $r_1 r_2'$, etc., to intersect the concentric arcs at 1_0, 2_0, 3_0, 4_0, 5_0, 1_1, 2_1, 3_1, 4_1, 5_1, etc.; at these intersections the graduation directly reads as far as one-sixth of a degree—namely 10′. This graduation is made not only for the whole 180°, but also for an extra 5° at either end of the semi-circle—that is, for 190°. The instrument has brass legs on the back at three points corresponding to P_1, P_2, and P, which may be tightly inserted into the wooden plate; and at Q_1 on the back a brass nail is attached, and from this point the line Q_1Q_2 is drawn at right angles to the line AB, provided for convenience in case this instrument should be used as a substitute for vertical circle.

The brass case for the magnetic compass is 106 mm. in diameter, and 21 mm. in depth, and is fixed by screws in the circular cavity G at the centre of the semi-circle. At S, the centre of the case and situated on CD, the radius giving the graduation of 90°, a steel needle is planted as a pivot; a magnetic needle about 100 mm. in length with a quartz cap is mounted on the pivot; and then the case is covered with a glass plate. At R and R' on the semi-circle where it is in contact with the rim of the case, two small holes are bored. These two holes are strictly on the line CD; and the fine thread, which is tightly strung through them, is used to adjust the orientation of the theodolite by making it coincide with the direction of the magnetic needle.

The alidades are attached to the brass arm KH, which may revolve about C, the centre of the graduation circles. The arm not only serves as a base of the alidades, but also as an index for taking the reading of azimuth; it is about 190 mm. long, from 18 mm. to 40 mm. wide, and 3 mm. thick. At the position corresponding to the centre of revolution a hole about 5 mm. in diameter is bored and pierced by a brass axis which has pierced the hole of the same size bored at the centre of the semi-circle, and the arm revolves smoothly. The alidades stand at 14 mm. and 124 mm. from the axis, facing each other 110 mm. apart. Their construction is almost similar to those belonging to the small magnetic theodolite; consisting of a pair of brass plates 60 mm. in height, 6 mm. in width, and 1 mm. in thickness, in each of which a slit 40 mm. in length, and about 0·3 mm. in width is cut; and when unused, the alidades are folded inside. On the arm and along the line connecting the two slits in the alidades there are two small holes *N* and *N′* 10 mm. long and about 2 mm. wide, which seem to have been used in adjusting the radius *CD* to coincide with the magnetic meridian, as in the case of the fine line *RR′*.

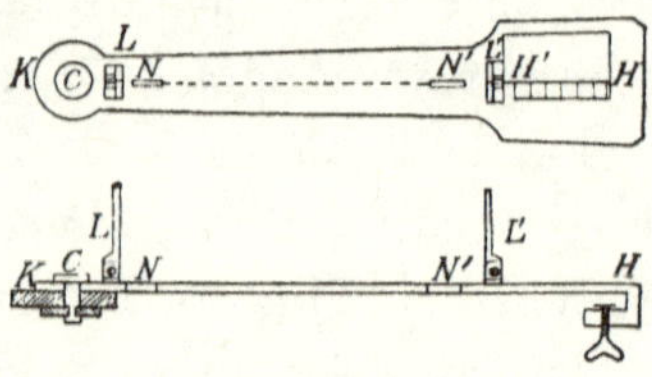

Fig. 9.

Towards the end of the arm there is a rectangular hole 33 mm. long and 15 mm. wide. *HH′*, one of its side-lines, strictly coincides with *CLL′*, the line connecting the slits in the two alidades and the centre of the axis. This line *HH′* forms the index in reading azimuth; in the form of a knife-edge it touches the graduated circle; and by the intersection of its edge and the diagonals, fractions of 10′ may be read by the eye-estimation as far as 1′ or 2′. The extreme end of the arm is bent around the edge of the circle to the back side where a clamp screw is provided for fixing the index at a desired point. No tangent screw is provided.

The above constitutes the main part of the semi-circular magnetic theodolite and weighs about 2 kg. It is fitted in a semi-circular zelkova-wood plate about 145 mm. in radius and about 25 mm. in thickness, on the back side of which a small hole about 9 mm. in diameter is bored at the centre. Into this hole the projecting piece of the stand is securely fixed.

The stand which was used in the field-work has not survived, but according to various records, it was constructed as follows: a wooden post was erected upon a tripod; and on the top of the post was fixed a square wooden plate with a small projection to fit the small hole made in the base plate of the theodolite. Moreover, it seems that a plumb-line or a plumb-needle was attached to the wooden post to secure the horizontality of the square plate.

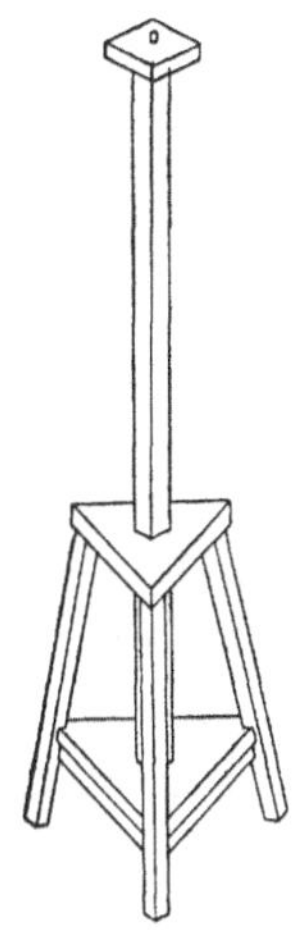

Fig. 10.

In using this instrument in field-work, the stand is first set up on the station and is so adjusted that the wooden plate on its top is horizontal; the magnetic theodolite is placed on the square plate so that the projection of the square plate fits into the hole on the back side of the base plate; and with the projection as its axis, the whole theodolite is slowly turned round upon the plate until the fine line RR' takes the identical direction with the magnetic needle. The instrument having been thus arranged, only the index arm, on which the alidades stand, is revolved till it points to the desired object, which is now sighted through the two slits in the alidades; the index arm is fixed at that position with the clamp screw and reading is taken by the index. With the completion of the measurement of an object, the clamp screw on the index arm is unfastened, and then the arm is slowly turned round to carry out the measurement of the azimuth of another object. As this instrument is graduated only for a semi-circle, the index arm cannot be revolved to all points on the whole circumference of

360°. Therefore, in setting up the instrument, if that part of the instrument which forms the arc is positioned to the south of *S*, *L* serves as the eye-alidade in sighting an object lying between east and west through south, and *L′* serves as the eye-alidade in sighting an object lying between west and east through north.

In spite of its minute graduation, the semi-circular magnetic theodolite is so simply constructed in other parts that it is liable to various instrumental errors. Especially the error caused by its eccentricity is unavoidable; and it is by no means a well-devised measuring-instrument. In practical use, however, this instrument revealed no appreciable shortcomings in comparison with other instruments, but displayed its own advantages in the greater minuteness of graduation as compared with that of the magnetic compass mounted on a stick, and also in the greater ease of transportation and handling than other magnetic theodolites of larger sizes.

Since the first tentative adoption of this instrument in 1803 in measuring the azimuths of mountains and islands, it was gradually favoured by the members of the surveying party; and in later years, this instrument alone was used in place of all other ordinary magnetic theodolites of larger sizes.

(*H*) *The Magnetic Theodolites of Middle and Large Sizes*

The large and middle-sized magnetic theodolites were the most accurate of all the magnetic theodolites used by Inō in measuring azimuth. Unfortunately, these instruments have not been preserved and their construction cannot be described so definitely as other magnetic theodolites. However, from a few documents and also from the construction of the instruments which are alleged to have been constructed chiefly after them, their construction may be fairly inferred.

A synthetic inference from all these extant materials shows that Inō's large magnetic theodolite was constructed as follows: on a circular wooden plate, probably of zelkova or Japanese cypress, 75 cm. in diameter, and about 5 cm. in thickness, a brass plate about 3 mm. in thickness is fastened with nails; the circumference of the brass plate is graduated by the diagonal scale so minutely that minutes can be read distinctly; a circular hole of a considerable size is made at the centre of the graduated plate to receive the axis of the index arm; beside this there is another circular shallow hole, *EF*, to receive the magnetic compass; the index arm consists of a brass plate about 75 cm. long, about 6 cm. wide, and about 5 mm. thick, and at its centre is attached an axis which smoothly revolves in the circular hole made in the graduated plate; and in the axis is bored a square hole to receive the lower end of the vertical axis, *QC*, supporting the telescope. Along the middle of the index arm a line *HCH′* which serves as the index line is drawn; both the ends of the index arm which reach over the graduated part of the circle are wrought into the form of a knife-edge along the line *HCH′*; and they are utilized in reading azimuth in the same manner as in the case of the semi-circular magnetic theodolite. At two points *L* and *L′* on the line *HCH′*, alidades for the naked eye are provided like those of the semi-circular magnetic theodolite; and at *N* and *N′* on the line *HCH′*, two holes are bored for sighting the magnetic needle while

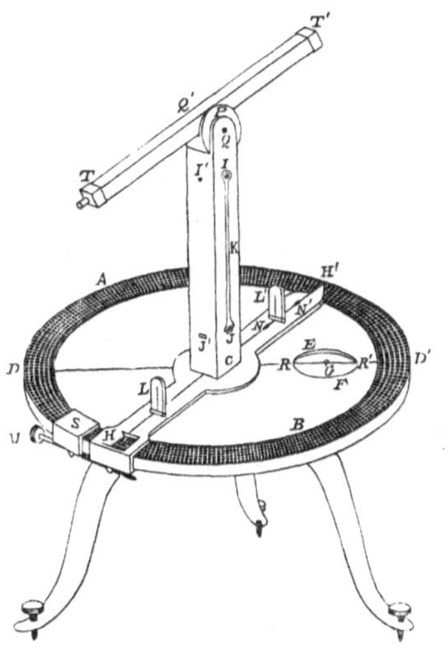

Fig. 11.

setting up the instrument in proper orientation. For fixing the index arm to the graduated circle a brass piece S with a clamp screw is provided; in order to give a minute motion to the index arm a tangent screw U which connects the piece S and the index arm is also provided. The vertical axis supporting the telescope is a wooden piece about 6 cm. square and about 60 cm. in height; the lower end is planted in the square hole made in the axis of the index arm, so that the vertical axis is rotated with the index; for testing the verticality of the axis, a plumb-needle K is to be hung from either of the two nails I and I' in the form of a knife-edge driven into two sides of it, and below the nails, indexes J and J' are attached. The upper end of the axis is concave in order to receive the wooden piece P attached to the barrel of the telescope; the telescope is attached to the vertical axis by a horizontal axis QQ' at right angles to the line HCH', so that the telescope may move in the vertical plane containing the line HCH'. The telescope is composed of lenses fitted in a square wooden barrel about 3 cm. square and about 60 cm. in length; the objective is a simple lens about 2 cm. in diameter and of a focal length of about 50 cm.; at the position of the principal focus a cross of silk fibre is provided; and the eye-piece is composed of two simple lenses. A magnetic compass of the same kind that is adopted in other magnetic theodolites is fitted in the shallow circular hole EF bored in the graduated circle and this is covered with a glass plate. The steel pivot G on which the magnetic needle hangs is positioned to be on the meridian DCD' marked on the graduated circle; and on the magnetic compass a fine line RR' is strung in the direction identical with the line DCD' and this is used in adjusting the instrument in the correct direction. The stand to support this instrument was a brass tripod whose legs no doubt had screws by which the graduated circle was adjusted to keep a horizontal position. It is said that the weight of this theodolite including the box was 30 kg., and that of the tripod and its box amounted to 35 kg..

The so-called middle-sized magnetic theodolite weighed 2·5 kg., its graduated circle was 36 cm. in diameter, and the other parts were smaller accordingly. Its construction was perhaps somewhat simpler, though more or less similar in style to the large magnetic theodolite. The stand for it was probably a wooden one.

The method of using these large and middle-sized magnetic theodolites was similar to that of the magnetic theodolites commonly adopted in the present land-surveying. This renders special explanations unnecessary.

Although Inō possessed the middle-sized magnetic theodolite prior to his undertaking of the survey of Yezo, the large magnetic theodolite was specially constructed for the purpose of measuring the accurate azimuths of distant mountains and islands only when the proposal of the survey of Yezo was nearly accepted by the Shogunate. Inō took this instrument to Hakodate. Difficulty of transportation, however, compelled him to leave it there and he was unable to use it in his survey of the south-eastern coast of Yezo. He scarcely used it in his survey-work up to 1803, but after 1804, as more convenience was offered him, he carried this instrument with him in his survey of the Chūgoku and Shikoku districts. It seems that he did not take it to Kyūshū. Inō's neglect of this instrument was no doubt because of the difficulty of transporting it. However, it was also due to the fact that in spite of the minutely graduated circle and telescope of the instrument, it had no adequate device for minute adjustment, and especially it had always to be orientated solely by a comparatively small magnetic compass, and, therefore, so many instrumental errors accumulated that the desired accuracy was considerably decreased; after all, it failed to prove a good deal more efficient than the other lighter and handier magnetic theodolites.

(*l*) *The Small Quadrant as Inclinometer*

As already stated, Inō used a small quadrant for measuring the inclination of the surveyed line, except during the first two years of his survey-work. According to the documents, he is said to have made it a rule to use an instrument about 20 cm. in radius, but it seems that size was not absolutely uniform. An extant instrument which is believed to have been used for this purpose is about 40 cm. in radius and is constructed as follows:—

On a quadrantal frame of zelkova 2 cm. in thickness and 40 cm. in radius, a brass plate of a similar form, 1·5 mm. in thickness is fixed; along the arc 39 cm. in radius, it is graduated in an ordinary way from 0° to 90° at an interval of 20′. At the centre *C* of the graduated circle is a brass axis 1·5 mm. in diameter, on which the index arm for reading the graduation is suspended. Parallel to *CA*, the radial line for 90°, a circular hole *LL′*, 9 mm. in diameter, as a sighting hole, is bored along the wooden frame. It is probable that either a cross-wire was provided in the hole, or some kind of tube with a still smaller hole was inserted in it as an aid in determining the sighting direction accurately; but now nothing remains but a round hole. At the approximate centre of gravity on the radial line *CD* for 45°, a round hole *E*, 6 mm. in diameter, is bored at right angles to the plane of graduation; a horizontal axis is inserted through the hole to fix the quadrant to a vertical pillar and also to enable the quadrant to take any

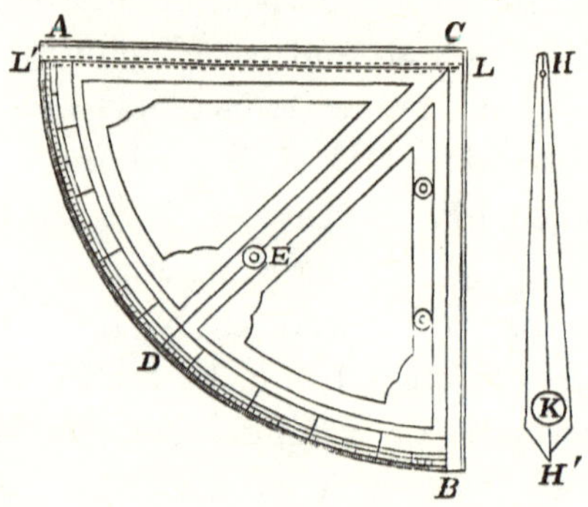

Fig. 12.

position within the vertical plane. The index arm consists of a sword-shaped brass plate 40 cm. in length, from 15 mm. to 30 mm. in width, and 1·5 mm. in thickness; at its upper end *H*, an elliptic hole is bored for suspending it on the axis at *C*; the upper edge of the elliptic hole is made in the form of a knife-edge so as to minimize its friction at the axis; and towards the lower end a brass weight *K* 30 mm. in diameter and 3 mm. in thickness is firmly attached.

The stand on which this instrument was set has been lost. It may be inferred, however, that the main body was a pillar with a horizontal axis to which the quadrant was fixed, and its lower end was sharpened to a conical form so as to be planted in the ground or in the tripod on which the small magnetic theodolite was erected.

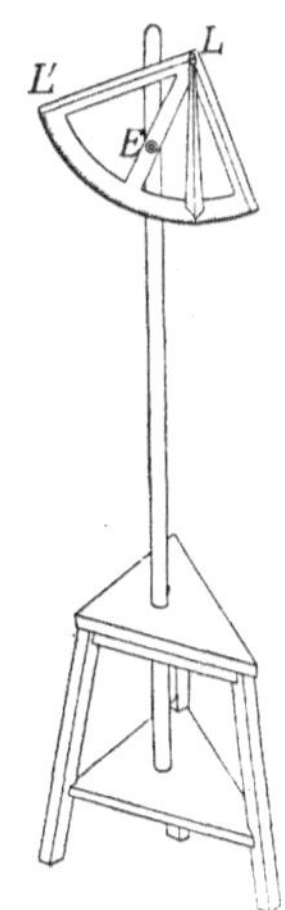

Fig. 13.

When Inō used this instrument, he first eliminated the index error by adjusting the position of the cross-wires in the sighting hole, making use of such a thing as the water surface of a pond. For actual determination of inclination, care was taken that the height of the sight mark be equal to that of the sighting hole of the quadrant. However, Inō's purpose of measuring the inclination of the route in his survey of the whole country was not to define the altitude of land, but simply to compute the horizontal distance of the route. Besides, the highways and the coast along which Inō carried on his survey were not much inclined, except at special points; in ordinary cases, therefore, no such elaborate process as to make the height of the sight mark exactly equal to that of the quadrant at each station needed to be adopted; a far simpler method was followed—namely, the instrument was so set up that the sighting hole of the quadrant was at the height of the eyes of the observer as he stood erect, and the eyes of the coolie of nearly equal stature standing at the station ahead were sighted.

In this manner, when measuring the inclination of the surveyed route, Inō made a simple observation with a small quadrant of simple construction; but in measuring the altitudes of mountains, he made an accurate observation with the above-mentioned quadrant or a large quadrant constructed for astronomical observation. However, the measurement of the altitudes of mountains was prompted by no particular necessity, but probably by his own curiosity; nor were many mountains measured. He chiefly measured the altitudes of some well-known mountains for computing their vertical height by making use of the horizontal distance to their summits determined by the method of intersection.

In his survey of the country Inō never practised precise levelling, but he eagerly investigated the proper method of practising it, because he keenly realized the necessity of precise levelling in some case, especially in opening a new water-way. The water-level-stand that had been handed down from an early date being inconvenient and unfit for practical use, Inō advocated the greater advantage of using either a quadrant or an instrument called *heijungi* (a levelling-instrument) especially constructed for this purpose. The quadrant he used for such precise levelling, however, was different from that used in measuring the inclination of the surveyed line; it was a minutely graduated instrument equipped with a telescope, tangent screws, etc., and fixed to a solid stand; that is, he temporarily adopted the instrument originally constructed for astronomical use.

The *heijungi* was a pure levelling-instrument, which has been lost, except for what are considered fragments of a miniature *heijungi*; so it is impossible to give the accurate details of the instrument. It may be roughly inferred from the "Ryōchidenshūroku." The principal part was a frame consisting of four pieces of hard wood or metal plate, *AB*, *DE*, *AE*, and *BE* as represented in the diagram. The line *AB* drawn along the piece *AB* and the line *DE* drawn along the piece *DE* meet at right angles. At *L* and *L′* on the line *AB*, two alidades for the naked-eye observation are attached, and a telescope *TT′* is fastened on the piece *AB* by screws *H* and *K*. By means of the screws *H* and *K*, the telescope may be so adjusted as to have its sight line parallel to the line *AB*. Towards the upper end of the piece *DE* is bored a circular hole *C* through which a horizontal axis is inserted to the vertical pillar, and which enables the instrument to turn on the axis; and at *F* towards the lower end a screw *N* is provided which may fasten the instrument to the pillar or give some slight movement to it. At the upper end *D* of the piece *DE* is planted a small needle on which a fine line with a weight is hung as a standard for adjusting the line *DE* to verticality. In order to prevent the weight from vibrating, and to secure its speedy stabilization, it is immersed in the water-holder *M*. The vertical pillar is erected by the tripod and may rotate about itself. As compared with the spirit-level instrument which is generally used to-day, the only difference is that this depended upon the plumb-line for adjusting the instrument. The method of levelling is quite similar in substance.

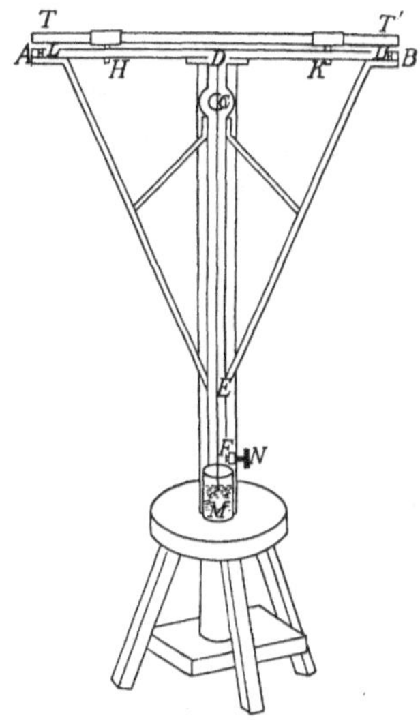

Fig. 14.

3 His Field-Note Reductions

The various books containing the results of reductions in

connection with the original notes of Inō's survey of Japan with the above-mentioned instruments have been lost or destroyed by fire, to say nothing of the original notes themselves; only a few fragmentary records are extant. However, from these few fragments of his computations and also from the pages of the "Ryōchidenshūroku," we may infer the method by which Inō reduced the materials obtained in his survey.

Inō consulted his field-notes every day and compared the azimuths for each segment of the surveyed route observed from both directions, direct and reversed; examined whether a gross error or mistake was introduced in reading the azimuths, and also in entering the distances by comparing them with the figures on the cards used in field-work. After confirming the absence of such error, he made the following computations.

(1) Taking the mean of the results of fore-sighting and back-sighting, the azimuth of each segment of the surveyed line was determined.

(2) By the result of testing the measuring-rope daily, the length of each segment of the surveyed route covered during that day was corrected accordingly.

(3) Where the route inclined, the length was multiplied by the cosine of the inclination to compute the horizontal length.

(4) If the width of a river had been surveyed by the method of intersection, its length was calculated according to the principles of plane trigonometry.

(5) If the azimuth of an object near the surveyed route had been observed with more than two magnetic theodolites, its value was determined by taking their mean.

All these items except (4) were readily computed directly on the pages of the field-notes. The azimuth and horizontal distance of each segment of the surveyed route thus reduced on the pages of the field-notes which were fundamental materials for map-construction, were extracted on another note-book whenever the field-notes proved inconvenient for synopsis. The azimuths of prominent mountains and islands especially observed with various magnetic theodolites from convenient points on the surveyed route being important materials for determining the positions of these objects and also for correcting maps, the results of such observation only were selected and copied in a separate book. This is termed the "Santō-hōiki" (The Record of the Azimuths of Mountains and Islands).

The materials thus arranged and copied were then computed as follows:—

(6) The horizontal length of each segment of the surveyed route was multiplied respectively by the sine and cosine of the azimuth to compute the component lengths in the directions of East-West and of North-South.

(7) The horizontal length of each segment of the surveyed route was summed up for a suitable section (usually the distance a survey-party covers in a day—namely, from 4 km. to 8 km.), and the E.W. component and N.S. component of the horizontal length of each segment were respectively summed up also for that section, to determine the E.W. component and N.S. component of the horizontal distance between the two ends of the section.

These computations were carried out not with the real *ri, chō,* and *ken,* but with what was represented on a suitable reduced scale (usually of 1:36000, or 1 *bu* for 1 *chō*). The computation of (7) shows that Inō treated the area containing one section simply as a plane, not as a spherical surface.

While travelling, Inō made it a rule, if possible, to reduce each day the survey-materials obtained during that day by the above-mentioned process, and to draw a map-draft on which he inquired whether errors had been introduced in surveying or not, lest doubts should arise in the future.

Then, in order to make it possible to compare the result of land-surveying with that of astronomical observation, with a view to correcting the former, the following computations were performed:—

(8) The small section covering several kilometers as defined under the preceding item was connected with others to compose a large section with an astronomical station at either end; the E.W. components and N.S. components of the small sections composing a large section were respectively summed up and were called the E.W. component and N.S. component, with which one end of the large section deviated from the other end. The N.S. components of large sections thus obtained were again summed up successively from the standard point in Yedo and the so-called total N.S. component was determined.

(9) The total N.S. component for each astronomical station was divided by 28·2 *ri*, or the length of one meridian degree as Inō had figured it out, and the difference of geodetic latitude between each station and the standard point was computed.

The computation of (8) here given shows that Inō treated the large section composed of small ones still as being on a plane. However, the N.S. component of the large section thus obtained practically gave a meridional length corresponding to the latitude difference between the two terminal points. Consequently the method of computation under (9) admits of no objection. On the contrary, the E.W. component of a large section obtained by summing up the E.W. component of each segment of the surveyed route is exceedingly indefinite as to its significance; as the earth is not really a plane, it is evident that this should not be called the E.W. distance between the two points. That Inō computed the so-called E.W. component for large sections which had no definite significance came to be a source of confusion in drawing maps.

4 His Method of Astronomical Observations

As already stated in the previous pages, Inō's knowledge of astronomy was simply what he had learned of that which his predecessors, such as Yasuaki Asada, Yoshitoki Takahashi, and Shigetomi Hazama, had cultivated with tremendous efforts. As to his method of observation or his measuring-instruments, he simply adopted what his predecessors had acquired after much consideration; and as he began a survey of the coast-line, he applied the various methods which he had practised and mastered to his determination of longitudes and latitudes of various stations. His process and method of measuring longitude and latitude in the field are not recorded concretely in any book; however, they may be fairly accurately given because of the quantity of the extant books and instruments relative to the subject and the documents and records handed down from the Astronomical Bureau at Asakusa.

The determination of latitude holds the most important position in all the field of the astronomical observation Inō always conducted personally during his survey of the whole country. His success in surveying vast areas without any great error is largely due to his achievement in measuring latitude.

His method of latitude measurement might be termed a kind of relative measurement: the meridian altitudes of a number of fixed stars were observed with the large accurate quadrant set up in the yard of his inn or at other suitable points every night during his journeys,—this result was compared with that of the meridian altitudes of the same fixed stars which had been accurately measured, previous to this, at the standard station in Kuroye-chō, Fukagawa, Yedo; and the latitude difference between the standard station in Yedo and the field-station in question was computed, and then this was added to or subtracted from the latitude of the standard station in order to obtain the latitude of the field-station.

It is notable that Inō, in his determination of latitude, did not rely upon the table of the declination of fixed stars which had been introduced by the I-hsiang-k'ao-ch'êng. And because he put the foundation of astronomical computation on his own observations, the measurement of the meridian altitudes of fixed stars at the standard station and the absolute measurement of latitude of the standard station became a task of supreme importance. The instruments and method adopted at the standard station in measuring the meridian altitudes of fixed stars were exactly similar to those adopted elsewhere; only at a field-station each of the fixed stars was observed as a rule not more than once, while at the standard station each was repeatedly observed for several nights or several dozen nights with a view to minimizing the error of observation. The absolute value of latitude of the standard station was based on the result of various observations; however, its principal material consisted of the result of the measurement of the altitudes of the circumpolar stars, especially of Polaris, at upper and lower culminations, though somewhat modified by the result of the measurement of the meridian altitudes of the sun at or near the two solstices.

At a time when a practical device for transmitting time to a distant locality was not yet invented, and when a portable timekeeper with sufficient accuracy had not been introduced from abroad, it was next to impossible for Inō to determine accurately the longitudes of distant points by astronomical observation. But it was extremely desirable to measure the longitude astronomically, as a comparison of the astronomical longitude of each station with the geodetical would eliminate error in surveying to a certain extent and considerably increase the accuracy of his maps. To attain this purpose by means of special heavenly phenomena such as solar and lunar eclipses, and the eclipses and occultations of Jupiter's satellites, Inō while travelling took every opportunity to observe at a field-station such phenomena, sometimes at the cost of tremendous labour, and comparing the result with that of simultaneous observation carried out at the Astronomical Bureau at Asakusa in Yedo, he attempted to compute the longitude of the field-station.

The chief point of such observation was to measure by the local time of the field-station the moment of the commencement, the close, or a certain intermediate moment of the phenomenon. The first important step for this observation, therefore, was to determine the local time astronomically and to keep it accurately for a necessary duration. In order to achieve this, Inō set up a transit-instrument at the field-station and, for several days before and after an eclipse, observed the time of the meridian transit of the sun and the most prominent fixed stars, by which he determined the constant of a pendulum-clock, and then by means of that pendulum-clock measured the time of an eclipse in the local true solar time.

In observing solar or lunar eclipses, he used the eclipsemeter especially constructed for this purpose; he observed the sun or moon several dozen times noting the clock-face times corresponding to the four contacts and to other various phases; then using all these data obtained he deduced the final value of times corresponding to the four contacts and to the middle of the eclipse. As to the eclipse or occultation of Jupiter's satellites, various telescopes were used by a number of members of the survey-party to measure the instant when the same phenomenon took place; he made it a rule to take the mean of the results obtained by several observers in order to reduce the observation error.

Solar and lunar eclipses were observed by Inō from the beginning of his survey of the country. But the phenomena occur only a few times a year—not often enough to be of practical use in the determination of longitude. On the other hand, the eclipses and occultations of Jupiter's four satellites occur quite frequently; such phenomena may be observed from any place as often as five or six times a month, on the average, excepting a few dozen days when Jupiter approaches the sun. However, when Inō first undertook the survey of the country, the elements of the motion of Jupiter and its satellites were unknown in Japan; it was, therefore, impossible to prepare their ephemeris beforehand, and to utilize such phenomena in measuring longitude. When Lalande's Astronomia was obtained in 1803 by the Astronomical Bureau, the motion of Jupiter's satellites became so clear that their positions might be roughly computed. After 1805, Inō was able to predict the approximate time of the occurrence of these phenomena according to the ephemeris of the eclipses and occultations of the four satellites, which was especially prepared at the Astronomical Bureau, and he made a point for observing them. Especially during his survey of the Kyūshū district, he was exceedingly anxious to determine astronomically the longitudes of several stations at least in order to compare them with those he had obtained geodetically, and did his very best to carry out such observation, as far as circumstances permitted.

Now, as a preliminary to this task of determining longitude by means of such celestial phenomena, it cost him a great deal of time in determining the local time; and the labour was not to be compared with that of determining latitude. Besides, the computed times of the eclipses and occultations given in the ephemeris were not very accurate as yet; it was necessary for the observers to commence observation a long time before the predicted time, and sometimes they had to wait all night long in vain, owing to a mistake in the computation. This task gave no small strain to the members of the survey-party who had laboured and exhausted themselves in their geodetic work in the daytime. And the result shows that the phenomena—even the eclipses and occultations of Jupiter's satellites which occur most frequently—were surprisingly seldom observed simultaneously at a field-station and at the Astronomical Bureau in Yedo; and their observation could not be expected to be accurate from the very nature of such phenomena. Therefore, the scanty observation material which cost Inō and his men great labour during their journeys proved of no practical use. Thus, the determination of astronomical longitude was practically a failure.

The phenomena Inō utilized in the determination of longitude were limited to the solar and lunar eclipses and the eclipses and occultations of Jupiter's satellites. No attempt was made by him to determine longitude by making use of the occultation of fixed stars by the moon, the lunar distance of fixed stars, the time of the meridian transit of the moon, or the meridian altitude of the moon. It is not that Inō was ignorant of the possibility of utilizing these phenomena, but that neither an accurate lunar table nor a star table had been prepared by that time, and the instruments also were inconvenient and inaccurate for such observation.

5 His Astronomical Instruments: Their Construction and Use

The instruments Inō used directly or indirectly in measuring longitude and latitude were quite numerous. But those which gave material assistance to his geodetic enterprise were the quadrant, the transit-instrument, the pendulum-clock, the eclipse-meter, the astronomical telescope, and the gnomon.

(*A*) *The Quadrant*

A quadrant was an instrument used for measuring the altitude of celestial bodies. Its form was originally introduced from Europe at an early date, since the main part of an instrument constructed at the beginning of the 18th century at the informal request of the Shogun Yoshimune Tokugawa was nothing but a quadrant provided with a telescope. According to a document extant, the quadrant was a brass plate 1·2 m. in radius, graduated by a diagonal scale and fastened to a plate of Japanese cypress wood. The telescope had cross-wires of the form of double crosses at its focus, and was so installed in the quadrant that it might be fixed at any desired position on the face of the quadrant or moved by the most minute motion. The instrument was adjusted to its proper position by means of a plumb-line. Although the death of Yoshimune Tokugawa checked the progress of such instruments for some time, the advancement of astronomy among the people gradually prompted the construction of improved instruments. About 1770, Yasuaki Asada and Naojirō Kataoka used large quadrants in practical observation. Then Takahashi and Hazama devoted themselves to the improvement of astronomical instruments, and when the mechanicians they had trained commenced to construct instruments, their improvement was marked and accuracy considerably increased. The quadrant at the Astronomical Bureau in Yedo which was specially built on the occasion of the compilation of the Kwansei Calendar was made by Tōzaburō Toda, a mechanician of Kyōto who had been patronized and instructed by Hazama. This was 2 m. in radius, and was more solidly constructed and accurately graduated than any other previous quadrant. The instrument was built according to the plan mentioned in the Ling-t'ai-i-hsiang-chih imported from China, and all our later quadrants were modelled after this.

Prior to his undertaking of the survey of Japan in 1800, Inō possessed two quadrants, one 180 cm. in radius, and the other 115 cm. The larger had been constructed about the year 1796 under Hazama's supervision at the request of Inō; the smaller was planned for greater handiness when Inō undertook his survey-work; it is supposed that they were both constructed by Yagorō Ōno and his son Yasaburō, mechanicians of Yedo.

Both these quadrants without their stands are still preserved by the Inō family; their construction is almost alike. The smaller of the two being a valuable relic which Inō carried to each station in his survey of the country and employed in his measurement of latitude, its construction may be described here.

The frame of the quadrant is of solid wood and formed as shown in the diagram. The two straight pieces, *AC* and *BC*, meeting at right angles, and the pieces *CD* situated half-way between the two pieces, are of Japanese cypress wood of the finest fibres and are 7 cm. in width, 3·5 cm. in thickness, and 115 cm. in length. The arc-piece *ADB* is 3·5 cm. in thickness and from 11 cm. to 12 cm. in width, and is also made of Japanese cypress of the finest fibres, triply, so that the fibres of the wood are connected lengthwise and sidewise to prevent the piece from warping. The outer edge of the arc-piece for more than 2 cm. is shaved down to a thickness of about 1 cm. and to this part is fastened the clamp screw which is

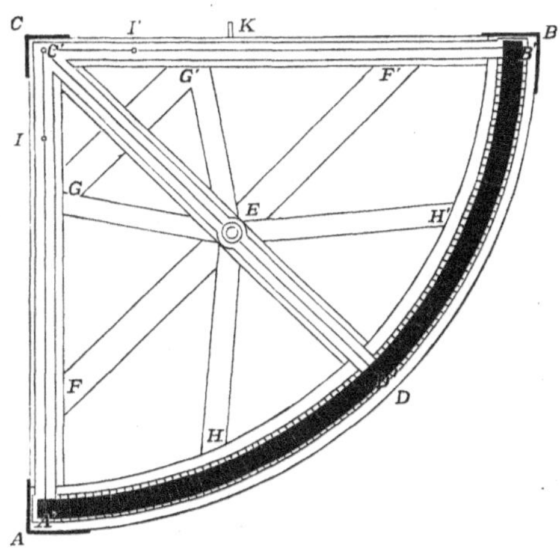

Fig. 15.

attached to the index arm. About 64 cm. from the angular point *C* on the piece *CD* a circular hole *E* about 2 cm. in diameter is bored for receiving an axis which, piercing through the hole, fastens the instrument to the pillar. *EG*, *EG'*, *EH*, and *EH'*, the four beams radiating from around the hole, and *GG'* and *FF'*, the two ties piercing the piece *CD* at right angles, intersect the pieces *AC*, *BC*, and *AB* in various directions and strengthen the whole framework in conjunction with the brass plates protecting the angular points, *A*, *B*, and *C*. To the three straight pieces *AC*, *BC*, and *CD*, and the arc-piece *ADB* which constitute the wooden frame, a brass plate about 6 mm. thick is fastened with more than 90 brass screws for graduation and strengthening the framework. The brass plates on the straight pieces are about 4 cm. in width and along their lengths are marked *C'B'*, *C'D'*, and *C'A'*, the fine lines representing the radii corresponding to 0°, 45°, and 90° respectively. The brass plate on the arc-pieces, which is more than 6 cm. in width, is graduated for the range of 92°—that is, 90° from *A'* to *B'* and an extra degree at either end; and a double reading of each degree is engraved, one reading from *A'* to *B'* and the other from *B'* to *A'* in the reverse order. The graduation which no doubt taxed the constructors greatest ingenuity is the most accurately cut of all the measuring-instruments of that day. The method of graduation is based on the diagonal scale which has been already given under the heading of the magnetic theodolite:—the 11 concentric arcs drawn with *C'* as their centre intersect the radial lines radiating from *C'* at an interval of 10′; then the diagonals which directly divide the interval to each 1′ make it possible to read by eye-estimation as far as 10″ or 5″. A part of the graduation is shown in Fig. 16; the radius of the innermost arc S_0 is 1052 mm., and that of the outermost arc C_{10} is larger by 41·2 mm.; the length of the arc for the angle of 10′ is 3·06 mm. at S_0 and 3·18 mm. at S_{10}.

Therefore, the dimension of the graduation-meshes is about equal to that of one *shaku* scale which has been already described; if compared in the matter of the dimension of graduation only, 1′ on the quadrant corresponds nearly to 0·001 *shaku*: sighting 0′·1 on the quadrant would compare with sighting 0·0001 *shaku* on the scale. However, it should be remembered that, while in marking 11 parallel lines on the *shaku* scale the uniformity of the intervals alone was important, the concentric arcs S_1, S_2,. . . S_9 on the quadrant could not intersect the diagonals drawn between the radial lines at such points strictly corresponding to the angles 1′, 2′,. . . .9′ respectively unless the radii of the concentric arcs be properly determined according to a definite law. This problem had been thoroughly investigated when the instrument was constructed, because the actual differences of the radii of the successive arcs satisfy the requirements with accuracy.

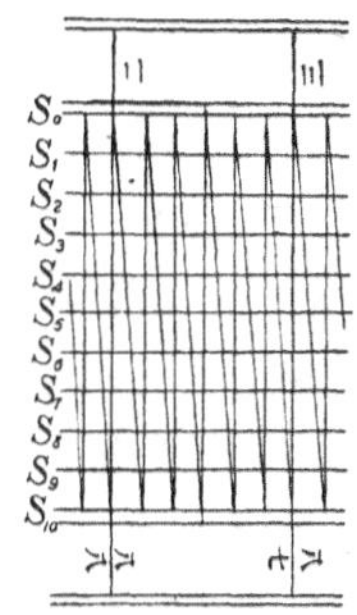

Fig. 16.

At C' the centre of the concentric arcs, a circular hole about 5 mm. in diameter is bored for receiving the rotation-axis of the telescope which is to be installed in the quadrant, and at I and I' on the lines $C'A'$ and $C'B'$, small holes about 4 mm. in diameter are bored. On the back of the piece AC a brass plate 4·5cm. in width,

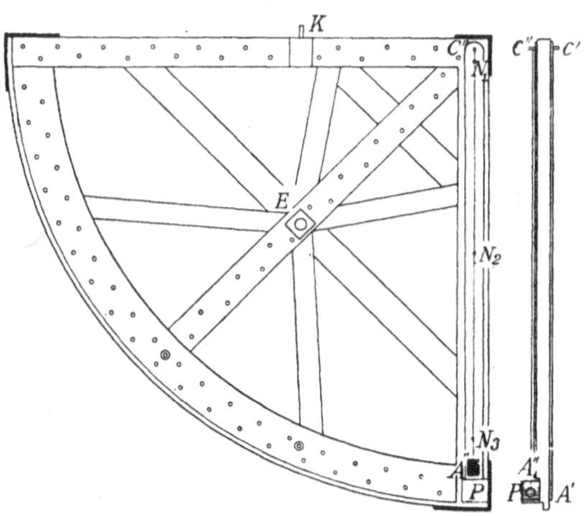

Fig. 17.

3 mm. in thickness, and about 110 cm. in length is attached. This plate has at its upper end C'' a circular hole about 5 mm. in diameter, similar to that at C' on the front face, and it is pierced by the axis of the telescope which has come through from the front face; the plate is fixed to the wooden frame by the screws N_1, N_2, and N_3, and along the length of the plate a straight line $C''A''$ strictly parallel to $C'A'$ on the front face is engraved. This last is used for adjusting the position of the instrument; that is, a plumb-line is suspended in front of $C''A''$ from the axis of the telescope, and the line $C''A''$ is so adjusted as to coincide with the plumb-line in setting up the instrument. The lower end of the brass plate is graduated by the diagonal scale as on the front face for the range of 1° 20′ having the line $C''A''$ in the middle; this arrangement was designed for the purpose of reading the amount of deviation of the instrument when it should lose its proper position during an observation and of correcting the results obtained. To the wooden frame below the graduation is attached a square brass case P, which is filled with water; the bob of the plumb-line suspended from the axis of the telescope is dipped into the water in order that its vibration may subside speedily; a slender wooden case, with glass windows, is fixed to the piece $C''A''$ covering the whole of the plumb-line to prevent it from wavering in the wind.

In order to fix the quadrant to the vertical axial pillar of the stand, a stout horizontal iron axis is applied in the circular hole at E on the piece CD, and is thrust through the circular hole bored in the axial pillar and then tightly fastened with a nut. The piece CB has an iron boss K, 3·5 cm. long and 1·5 cm. square protruding upwards at a position directly above the hole E; when the instrument is fixed to the pillar, this boss goes in the centre of three screws attached to the pillar and by the delicate motion of these screws the position of the quadrant is so adjusted that

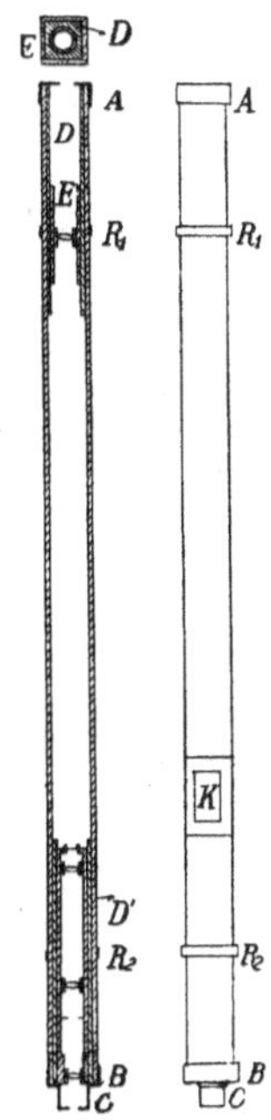

Fig. 18.

the line C' A' (the line $C''A''$, consequently) keeps an exact vertical direction.

The telescope attached to the quadrant, as shown in the diagram, consists of four simple double convex lenses arranged in a square wooden tube. The square tube is made of four pieces of paulownia timber about 4 cm. square in section and 117 cm. long, and two paulownia tubes 30 cm. long with circular holes about 3 cm. in diameter, D and D', are inserted into the square tube at either end to make its hole cylindrical. The outward surface of the square tube is varnished with black lacquer, and brass caps are applied at both ends, A and B; and at the points 17 cm. from A and 16 cm. from B, two brass belts, R_1 and R_2, strengthen the wooden tube. About 33 cm. from B, a rectangular glass window K is provided for the purpose of illuminating the cross-wires in the tube with candle-light from outside during night observation.

The objective of the telescope consists of a simple double convex lens 28 mm. in diameter with a focal length of about 71 cm. In fixing this in the telescope, the lens is first fastened by two ring-formed brass washers F and G within E, a brass cylinder 30 mm. in outer and 28 mm. in inner diameter, and 10 cm. in length; and E is inserted and fixed within D, the paulownia cylinder. This objective is not located near the terminus A, but about 18 cm. distant from it, being restrained by the focal length of the objective. In the brass cap at the terminus A, a circular hole 16 mm. in diameter is bored as the window, which limits the incoming light to the central part of the lens; a shutter is provided at the opening to close it when not in use.

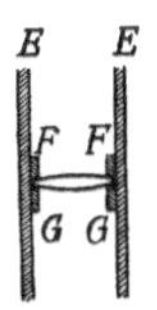

Fig. 19.

The eye-piece consists of three simple double convex lenses which are arranged and fastened almost in the same manner as the objective within a brass cylinder about 3 cm. in outer and about 28 mm. in inner diameter, and 30 cm. in length. The lens farthest from the eye is about 27 mm. in diameter, the other two being somewhat smaller; 12·8 cm. distant from the foremost lens, the second one is fastened; about 3·3 cm. from this, a diaphragm, and 12 cm., the third one, or the lens next to the eye; and the eye-end is covered with a brass cap which has C, a circular hole about 5 mm. in diameter. This brass cylinder serves not only in arranging and fastening the eye-piece, but also as a framework for the cross-wires which are strung at the foremost end of the cylinder, the position about 2·7 cm. ahead of the first lens. To string the cross-wires a thin brass ring with 4 small holes H_1, H_2, H_3, and H_4 is inserted and fastend to the end of the brass cylinder; a peg covered with paper is planted in each of the small holes; then the two ends of the silk fibre are wound round the pegs at H_1 and H_3 respectively and the line is strung by turning the pegs; and in a similar way another line is strung between H_2 and H_4 to make a cross. The brass cylinder fastening the eye-piece and the cross-wires is inserted into D', the paulownia cylinder, from the back end until the cross-wires take the most suitable position—namely, a position corresponding to the principal focus of the objective or about 71 cm. from the objective, when it is fixed by a screw driven from one side of the brass cap at the end B. It is impossible to adjust the eye-piece freely to one's particular eyesight, though, as the end of the screw is pressed into a groove J on the outside of the brass cylinder, the direction

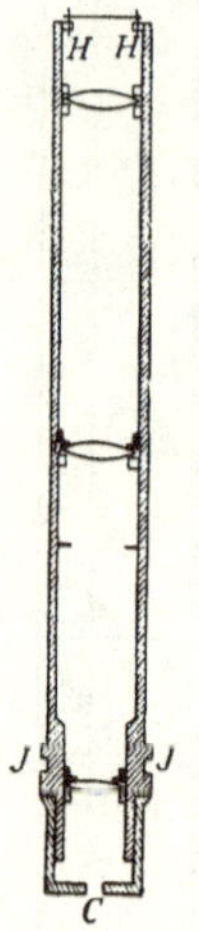

Fig. 20.

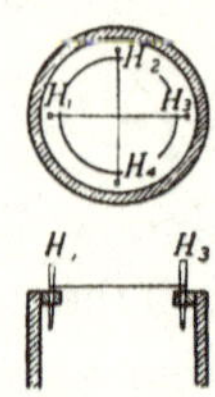

Fig. 21.

of the cross-wires may be altered freely if the screw is slightly loosened and the brass cylinder turned. In attaching the telescope to the quadrant, the cross-wires should be adjusted and fixed so that one of the wires takes the horizontal direction and the other the vertical.

In installing the telescope on the quadrant, it is first fixed to a solid wooden plate to be used as the index arm in reading the graduation. The index arm, consisting of a wooden plate 2 cm. in thickness, 9·5 cm. in width, and about 115 cm. in length, has a brass plate 3·5 cm. in width and 2 mm. in thickness nailed along one of its sides for the whole length. At C''' on the brass plate near one end, there is a hole about 5 mm. in diameter to receive the axis which has pierced C' of the quadrant and to become the axis of rotation for the index arm. The other end of the brass plate fits in a section of the graduated part of the quadrant when the index arm is attached to it, where the wooden plate is cut out and instead a brass index plate T for reading angles is screwed down. T is 2·5 cm. in width and 4·5 in length; one edge is in knife-edge form and in close contact with the graduated face and serves as the index line in reading angles.

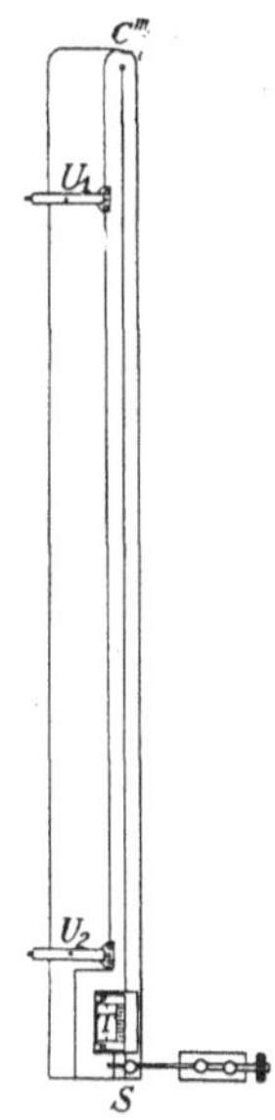

Fig. 22.

For the purpose of attaching the telescope to it, the index arm has two solid square brass frames, U_1 and U_2 nailed to it at two points, one about 17 cm. and the other 101 cm. from the end C'''. The square frames are about 7 cm. square in section and a screw is provided at the centre of each side. The tube of the telescope is now thrust in the square frame until the brass belts R_1 and R_2 on the tube reach U_1 and U_2 respectively; and then the screw on each side is suitably adjusted until the line of sight of the telescope is parallel with the index line $C'''S$ on the index arm; and at this position the telescope is firmly fixed.

The back of the end S of the index arm has a crooked brass plate, which goes round the edge of the quadrant when the arm is fixed to the quadrant. A screw provided on it clamps the index arm to any desired position of the quadrant face. And on the front face of the end S there is a boss provided with a female screw connected with a tangent screw which may give a minute motion to the index arm.

The axis of rotation for the index arm when it is installed in the quadrant is a steel bar about 5 mm. in diameter and about 9 cm. in length which exactly fits the two circular holes, C''' and C', and which is fixed on the front and the back by means of washers. At the part of the axis sticking out on the back side, a groove is cut around it in a ring form, which serves as a point of suspension of the plumb-line for determining the verticality of the quadrant. The plumb is a brass bob about 3 cm. in diameter; a fine thread about 110 cm. in length is fastened to it; and the other end of the thread is attached to a small piece of brass with a circular hole in it. By means of the circular hole the plumb is suspended on the axis; the fine thread hangs immediately in front of $C''A''$, the line marked upon the back of the quadrant; and the plumb dips in the water, filling the case P attached at the lower end of A''.

The stand on which the quadrant was fixed has been lost; but according to documents, it seems that it was so constructed that it might be taken apart when necessary. Fig. 23 will show how it appeared. The stand was of wood; the most important part or the axial pillar was 9·7 cm. square and 206 cm. in length; it had a circular hole E', 2 cm. in diameter which was protected with an iron washer, at a point about one-third of the whole length down from its upper end. About 47 cm. above the hole was provided K', a metal plate bent into a frame. The iron axis piercing the hole E of the quadrant was thrust into E', which in turn fastened the quadrant to the axial pillar by means of a nut. The iron boss K on the upper part of the quadrant was inserted in the frame K'; the position of the instrument was then adjusted by means of the screws attached to the frame. To either end of the axial pillar, conical steel pivots M_1 and M_2 were fastened so that M_1M_2 served as the axis of rotation for the axial pillar. The caps for the axial pillar were made of steel plates with proper cavities for receiving the pointed ends of the axial pillar, one at N_1 on the lower side of the arm and the other at N_2 of the base. The cap N_1 was fixed to the arm while N_2 was not fixed to the base, but lay under the two bent iron plates which were nailed down to the base. When the axial pillar was erected, N_2 was moved until the pillar stood vertical and then it was fastened to the base by means of set-screws.

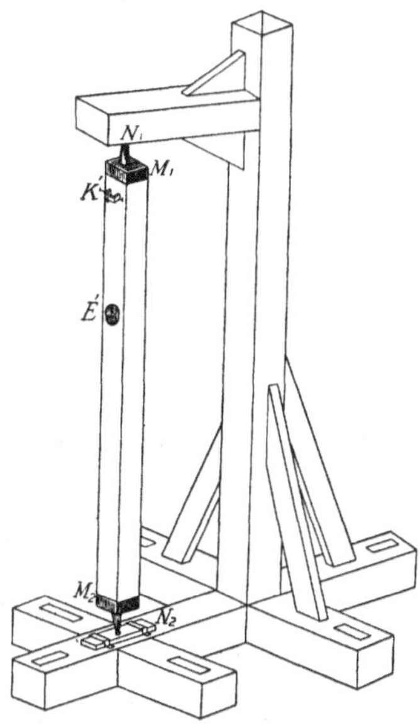

Fig. 23.

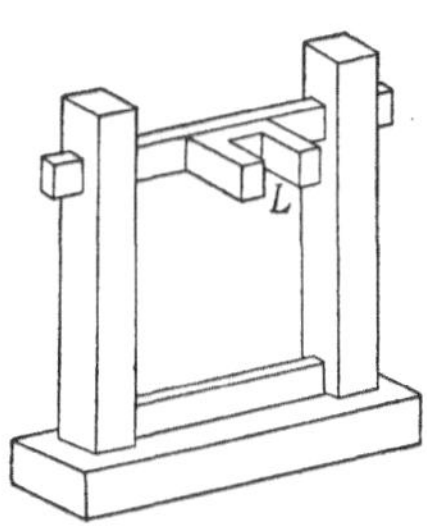

Fig. 24.

When Inō first designed this stand, it was so constructed as to be taken apart for convenience in transportation. In later years, however, his survey covered only 10 km. each day, and he was permitted to demand as many men and horses as he wanted. It thus appears that the stand was transported without being dismantled, except on special occasions. According to a document, there was also an extra wooden stand attached to the stand, as shown in Fig. 24. At L a wooden piece with a groove exactly fitting the vertical piece of the quadrant was provided for supporting a part of the vertical piece of the quadrant in a desired azimuth and for holding the instrument in that azimuth during an observation.

Though a quadrant may be used in measuring the altitudes of heavenly bodies or other objects in any azimuth, Inō used it chiefly in measuring the meridian altitudes of the fixed stars, sun, moon, and planets. Therefore, when the direction of the meridian was once determined by some method, Inō certainly had some means by which the direction was always kept and by which, no matter how often the quadrant was rotated on its vertical axis, it was possible to have the instrument return readily to the meridian plane and stand still. As the extra stand above-mentioned was very inadequate for such a purpose, Inō must have had some simple suitable design, though the method he adopted is not certainly known to us.

In setting up the quadrant, it was most important to secure the absolute verticality of the line $C'A'$ engraved on the piece CA of the instrument. In order to make such adjustment, first, the axial cap N_2 on the base of the stand was moved so as to make the axial pillar, and consequently the face of the quadrant fastened to its side, take a strictly vertical position; then by manipulating the screws attached to the frame K', the boss K on the top-piece of the quadrant was moved slowly on either side until the plumb-line suspended from C'' on the back side of the instrument coincided exactly with the engraved line $C''A''$; and then all the set-screws were fastened. Since in adjusting the line $C'A'$ to take the vertical direction, not the line $C'A'$ itself, but the line $C''A''$ was tested instead, it was important that $C''A''$ should first be made parallel with $C'A'$. And for this purpose, a small axis was thrust in the small hole bored at I on the line $C'A'$ and another plumb-line was suspended from this. As this plumb-line coincided with the line $C'A'$, it was examined whether the plumb-line on the back coincided with $C''A''$, and if not, adjustment was made by manipulating the screws, N_1, N_2, and N_3, attached to the brass plate on the back.

The sight line of the telescope and the line $C'''S$ marked on the index arm were adjusted parallel to each other in order to minimize the index error. But if the meridian altitudes of the fixed stars were measured on both the northern and southern sides, the amount of the index error could be computed easily and consequently the influence of the index error could be eliminated from the required value of the latitude; therefore, during the earlier period of his survey-work, Inō did not pay much attention to such adjustment, but quite often used the instrument without correcting even a considerable index error. In later years, however, he took the greatest care in adjustment, and did his best to minimize the instrumental error.

After adjusting the instrument, it was necessary to determine the direction of the meridian. The simplest methods of determining it by astronomical observation then adopted in Japan were to make use of the equal altitudes that a heavenly body occupies before and after it passes the meridian or to avail oneself of the elongation of the circumpolar stars. These methods, however, were impossible to apply to an instrument without a horizontal circle. Therefore, in determining the direction of the meridian with a quadrant, the following methods were adopted. The first method was to set up the instrument in the direction of the approximate meridian, by which he measured the time of the upper and lower culminations of a circumpolar star; and by comparing the time interval from the upper culmination to the lower with that from the lower culmination to the upper, the deviation of the temporary position to the true meridian was computed; from this result the temporary position was corrected; and then repeating the process the direction of the instrument was corrected successively until no deviation was detected. The second method was to observe the times at which any two fixed stars, whose difference of right ascensions had been accurately known, passed the assumed meridian, by which the deviation of the assumed meridian was computed and the assumed position corrected; a similar process was repeated several times for further correction until no deviation was detected. Of these two methods, the first, though not requiring a knowledge of the right ascension of the circumpolar star, took several long nights to set the instrument accurately; therefore, it was used only in such a case as to set a transit-instrument at a permanent station. It seems that Inō probably adopted the second method in setting the quadrant at field-stations, but because he was convinced that the magnetic meridian and the true meridian coincided, it is possible that, at field-stations, he more frequently determined the direction of the instrument by a magnetic compass.

In observing the meridian altitude of a fixed star, the vertical piece CA of the quadrant was turned to the south or north according to whether the star was in the south or north of the zenith, and the instrument was fixed in the meridian; it was examined whether the plumb-line suspended from C'' coincided with the engraved line $C''A''$; if not, it was strictly adjusted; then the telescope was directed to the assumed altitude of the star, and waited for it to come into the field of the telescope; when the star came, the telescope was slowly moved by the tangent screw so that the star proceeded along the horizontal line of the cross-wires; and as the star passed the vertical line of the cross-wires, it was so manipulated as to be precisely on the horizontal line of the cross-wires; the telescope was clamped in this position, and by the index on the index arm, the angle on the graduated circle was then read; and it was tested again as to whether the plumb-line really coincided with the line $C''A''$; if any change should arise, the deviation was read by the graduation attached to the part A'' and by this the angle which had been read was corrected. The reading of such angles was always taken with the naked eye, ordinarily down to 5″. Once observation was commenced, as many fixed stars as possible were observed one after another by this method; it seems that preparations for the observation of one star after finishing another were very speedily made; and the observation records show that two fixed stars with so small a difference as 30^s in right ascensions and yet differing in the direction of culminations, one in the south and the other in the north, were both measured as to their altitudes. It is evident that, in measuring the altitude of a fixed star, the observer had at least two assistants, one of whom informed the observer of the direction and the assumed altitude of the fixed star which was to culminate according to the star table, and reminded him of the time of culmination by the pendulum-clock, and who also adjusted the verticality at the same time; and the other assistant was to illuminate the cross-wires through the glass window of the telescope with a lantern and to illuminate the graduated face as the angle was read and also to do other jobs. Throughout the whole period of his surveying, Inō always was in charge of altitude observation except when hindered by illness or some unavoidable circumstances; and assistance was given him by his own pupils.

According to the various documents, the fixed stars Inō made use of in determining latitude were generally those larger than the fifth magnitude, chosen from the stars, the altitudes of which he had formerly measured accurately at his home station in Kuroye-chō, Fukagawa, and registered in his star table; and the observation was made usually according to the following rules.

(1) To make observation of as many fixed stars as possible.

(2) Not to make observation of the stars in only one direction of the zenith, but to observe the stars in both directions, south and north as equally as possible.

(3) To exclude the fixed stars with a small meridian altitude, but to choose and observe those of the largest possible altitudes.

And as to the first rule, Inō thought it sufficient to observe at an ordinary field-station from 15 to 25 stars in a single night; but at a station where an eclipse was to be observed or where an especially accurate latitude was required, observation was repeated several successive nights, and the stars observed in a single night numbered sometimes more than 70. As to the second rule, it is true that the natural inequality that stars culminating in the south in a definite time are more numerous than those culminating in the north could not be helped, but so far as weather permitted Inō made it a rule to observe the stars in both directions as equally as possible. As to the third rule, though he sometimes observed stars culminating in the south at such a low altitude as 22°, he usually measured those at more than 40°, and especially took every care not to overlook those culminating near the zenith.

Inō compiled the results of his observation of the meridian altitudes of fixed stars made at various stations, and these under the title of "Hokkyoku-kōdo-sokuryōki" (The Record of Latitude Measurements) numbered several dozen volumes. All have been lost, except one which records the results of observations made in 1802. In order to show the real state of the measurement of altitude, an example is here given.

Wakamatsu, Aidzu, Mutsu.				July, 26, 1802.
Star	Mag.	*N* or *S*	Meridian Altitude	Altitude Difference
γ Herculis	3·8	*S*	72° 5′ 10″	1° 49′ 50″
β Herculis	2·8	*S*	74 24 40	1 49 20
η Draconis	2·9	*N*	65 31 30	1 50 30
ζ Herculis	3·0	*S*	84 26 20	1 49 40
κ Ophiuchi	3·4	*S*	62 11 15	1 48 45
ε Urs. min.	4·4	*N*	45 9 12	1 49 42
α Herculis	3·3	*S*	67 6 15	1 49 15
ζ Draconis	3·2	*N*	61 32 32	1 50 32
λ Herculis	4·4	*S*	78 44 25	1 49 35
α Ophiuchi	2·1	*S*	65 12 30	1 48 30
ν^2 Draconis	4·8	*N*	72 10 35	1 50 35
β Ophiuchi	2·9	*S*	57 7 50	1 50 10
μ Herculis	3·5	*S*	80 19 50	1 48 40
ξ Draconis	3·9	*N*	70 33 20	1 50 20
γ Draconis	2·4	*N*	75 57 1	1 49 31
Map 37° 30′				

The fourth column of this table indicates the apparent altitudes read on the quadrant and adjusted for index error; the fifth column the differences between the apparent altitudes of the fixed stars given in the fourth column and those of the same stars observed at the fundamental station at Kuroye-chō in Yedo; the notation "Map 37° 30′" is the latitude of this locality, as adopted in his map as the final result. The magnitude of the stars is newly added for reference.

To compute the latitude of a given point from the apparent

meridian altitudes of fixed stars observed there, it is of course necessary to use a table of the declinations of fixed stars or some substitute for it. In Inō's day the position of the fixed stars was fairly accurately known from the I-hsiang-k'ao-ch'êng which had been compiled in China in the middle of the 18th century. The positions given in the book were those at the winter solstice of 1743, and the subsequent positions were to be computed simply by adding the term of precession in simple proportion, the numerical value of precession for the right ascension and declination being given in the book. However, as the constant of precession was quite inaccurate and the method of computation crude, the calculated place for about 1800, or when over a half-century had elapsed since the compilation of the table, differed considerably from the true place. By this reason, Yoshitoki Takahashi and others, when they were appointed official astronomers, intending to determine the positions of fixed stars for themselves and compile a star table, began to measure the apparent altitudes of the more prominent stars; and based upon such observation from January in 1801 to March in 1803, a table of the apparent meridian altitudes of 211 stars at the winter solstice of 1801 was made by them. About this time Inō also was engaged in similar observation at his residence at Kuroye-chō in Fukagawa and constructed a star table of his own, to which he referred in computing the latitude of a field-station by an observation of the meridian altitudes of fixed stars made at such a field-station. The number of the fixed stars registered in his table in 1803 was about 350, and the table he compiled in 1813 really contained not less than 500.

The method of observing the meridian altitudes of fixed stars at the base station was exactly similar to that adopted at any field-station; only the observation of one fixed star was repeated at least more than ten times in order to eliminate observation error. All the star tables compiled by Inō contained, not the altitudes read on practical observation, but those corrected not only for such instrumental errors as error in verticality or index error, but also corrected for precession—namely, the apparent altitudes of the fixed stars reduced to a certain definite epoch, though including an error arising from atmospheric refraction. The perfection of a star table, however, was not the original purpose of his compilation of the table; he only desired to use it for the determination of the latitudes of various field-stations. Therefore, in the earlier years of his survey-work, Inō's method was to observe in Yedo before and after each journey the altitudes of only those fixed stars which he could utilize at field-stations during that journey, and to compare the results with those obtained by observation at each field-station. With the development of his survey-work, however, the period of his journey lasted for a whole year, the number of the required stars increased, and the materials of the fundamental observation in Yedo also accumulated until the star table came to be a bulky one.

In making the fundamental observation at the home station in Yedo, Inō took pains to render the observation conditions as nearly similar as possible to those under which observation was made at the field-stations. It seems that, in order to obtain full effect of the so-called relative measurement, he observed at the home station also with the quadrant, 115 cm. in radius, which he carried in his journeys. The records of observation concerning this fundamental observation have all been lost. Only a brief record of observation made from Nov. 19th, 1802 to Jan. 5th, 1803 is preserved.

Of all the star tables compiled on the basis of the results of such fundamental observation, only two original documents are extant: one containing the results of observation in 1802 and 1803, and the other giving the positions at the winter solstice of

1812. To give an idea as to the form of its contents, a portion of the latter is here reproduced.

Star Places at the Winter Solstice of 1812				
Name of Star	Mag.	Observed Altitude	Calculated Altitude	Calculated Right Ascension
o Herculis	3·8	*S* 83° 5′ 0″	83° 4′ 4″	0° 3′ 38″
μ Sagittarii	4·0	*S* 33 15 30	33 13 25	0 40 29
η Sagitarii	3·1	*S* 17 35 30	17 32 59	1 18 19
δ Sagitarii	2·8	*S* 24 28 0	24 26 3	2 15 5
ε Sagitarii	1·9	*S* 19 55 10	19 52 6	2 57 38
η Serpentis	3·4	*S* 51 24 40	51 24 28	3 2 33
α Urs. min. (80° after L.C.)	2·1	*N* (35 23 0)		3 30 16
λ Sagittarii	2·9	*S* 28 51 40	28 49 39	4 7 34
φ Draconis	4·2	*N* 54 26 30	54 26 40	5 51 0
χ Draconis	3·7	*N* 53 1 40	53 1 4	5 59 5
α Lyrae	0·1	*N* 87 3 0	87 3 49	7 38 24
δ Urs. min.	4·4	*N* 39 6 30	39 6 4	8 21 40
φ Sagittarii	3·3	*S* 27 11 0	27 9 17	8 29 54
....	....	....	....	....

The third column of this table gives the apparent altitude determined by observation at the home station—namely, the altitude including atmospheric refraction. The fourth column indicates the calculated altitude which was obtained by correcting the declination given in the I-hsiang-k'ao-ch'êng to the value at the winter solstice of 1812 applying the precessional term by simple proportion and then by combining with the latitude of the fundamental station at Kuroye-chō in Fukagawa which he determined to be 35° 40′ 30″*N.*; and this contains no refraction error. The fifth column shows the calculated right ascension at this time, which was obtained by correcting the right ascension given in the I-hsiang-k'ao-ch'êng, and the point of winter solstice is taken to be 0°. The altitude of a star like Polaris, given in parentheses, is that when out of the meridian; this row gives the apparent altitude at the hour angle of 80° after the lower culmination. The magnitude of the fixed stars given in the second column is newly given for reference; there is no such column in the original document.

At the end of the table Inō prudentially reproduces a table of correction for atmospheric refraction which was introduced by the Sequel of the Li-hsiang-k'ao-ch'êng. But it seems that he did not give much credit to the values contained in the table.

Moreover, such phenomena as aberration and nutation came to be known among the Japanese when Yoshitoki Takahashi obtained the Astronomia by Lalande in 1803. Therefore, when Inō surveyed the eastern half of Japan, it is probable that no attention was paid to such phenomena. By the time he surveyed the western half, his knowledge in this direction had doubtless been much improved. However, because no mention is made of these questions in his books, it is evident that he failed to correct his measurement of the altitudes of the fixed stars for these phenomena even in his last years.

Again, in obtaining the latitude difference between any field-station and the fundamental station, the refraction error for the same fixed star at either point was supposed to be equal, and the difference was computed only by comparing the apparent meridian altitude of a fixed star observed at the field-station and that given in the star table based on the observations made at the fundamental station. In order to eliminate the index error of the quadrant, the latitude differences obtained from the stars culminating in the north and those from the southern stars were averaged separately; and the mean of these two average values was adopted. As an example, let us consider the observation made at the town of Wakamatsu which has already been referred to in a previous page:—

The mean of latitude difference obtained from
6 north stars . . . . 1° 50′ 12″
The mean of latitude difference obtained from
9 south stars . . . . 1° 49′ 18″
The mean of the above two . . . . 1° 49′ 45″

This value, 1° 49′ 45″, is the latitude difference between the town of Wakamatsu and the base in Yedo. By adding 35° 40′ 30″, or the latitude of Kuroye-chō in Yedo, 37° 30′ 15″ is obtained for the latitude of Wakamatsu. It is beyond doubt that, because of the omission of numerous corrections in the point of computation as already mentioned, this result could not be relied upon down to seconds, even though no error whatever was committed in observations. Inō seems to have considered an error less than 30″ as unavoidable in computations as well as in observations; and this result was compared with that obtained geodetically and the disposal of fractions of each 30″ being determined, the value with its lowest figure reduced to 30″ was adopted as the final latitude of the point, and was registered on the map and the "Enkai-jissoku-roku" (The Record of the Coast-Survey). In the example for the town of Wakamatsu, 37° 30′ 15″ was reduced to 37° 30′.

The above was Inō's method of determining latitude at an ordinary station; he made it a rule to observe for one night. At Mimmaya which is situated at the extreme north of the Main Island and is of the greatest importance in determining the length of the meridian degree of the earth, and at some other stations he considered important, astronomical observations were continued for several nights; and as to the computations of latitude differences, care was taken in retaining or omitting fractions, nor was the last result reduced to 30″, but retained even to 10″ or 5″ for further reference.

The absolute value of the latitude of the fundamental station in Yedo was determined, as already stated, chiefly from the upper and lower culminations of the circumpolar stars and the meridian altitudes of the sun at or near the two solstices, the former being observed with the quadrant. His method was to await the best season for each circumpolar star and, by the same method with which he observed the meridian altitude of an ordinary fixed star, to observe their upper and lower culminations repeatedly, in the evening and in the morning, for many days, to correct the result for the index error of the telescope and for atmospheric refraction and to take their mean. The stars Inō made most use of, it seems, were two—α Urs. min. and γ Cephei.

In his determination of the latitude of the fundamental station, Inō also referred to a pair or several pairs of fixed stars culminating in the south and north of the zenith at short intervals of time and at approximately equal altitudes: he observed their apparent altitudes and corrected them for atmospheric refraction, then by means of the declinations computed according to the star table, the values of the latitude were figured out from the southern and northern stars respectively; and their mean was the value he referred to. This may be considered something like the so-called Talcott's method: errors arising from the index error of the telescope, the imperfection of the graduation, the strain of the quadrant, and the inaccuracy of correction for atmospheric refraction, were almost completely eliminated and did not affect the final result, and naturally this method was considered one of the best practised in Inō's days. The absence of an accurate star table, however, made it impossible to regard the value obtained by this method very important in determining the latitude of the fundamental station.

The documents fully revealing the extent to which Inō made observations with the quadrant and the other instruments, or showing how he reduced and combined the results of his observations in determining the absolute value of the latitude of the base station have not been discovered. It is clear from various documents that as the confirmed latitude of the fundamental station at Kuroye-chō in Fukagawa a value of 35° 40′ 30″*N*. was consistently adopted. And it is probable that this value with no fractions of 30″ was obtained not naturally, but by a convenient disposal of fractions below 30″. As to the latitudes of Mimmaya, Kagoshima, and other points he considered most important, their original values, with no reduction of fractions, are discovered in various documents; on the other hand, the latitude of Kuroye-chō was always set as 35° 40′ 30″ even in the computations that demanded utmost accuracy. This fact will prove that the original value for the point, based on the materials of various observations, was no doubt quite close to 35° 40′ 30″.

Although the chief purpose of the quadrant was to measure the altitudes of various celestial bodies, Inō sometimes substituted it for the transit-instrument in determining local time. This was quite unusual at that time, for in measuring local time accurately the transit-instrument specially constructed for this particular purpose was used.

(*B*) *The Transit-Instrument*

Determination of the time at which heavenly bodies pass the meridian is naturally of great importance for several purposes. Even at the very beginning of the 18th century there existed some instruments with proper devices for such measurement; for instance, there was one constructed at the request of the Shogun Yoshimune Tokugawa. These instruments, however, were far from perfection and were inconvenient as to their adjustment and use; besides, on account of their complicated construction and tremendous size, they were beyond the reach of common astronomers. The so-called transit-instrument was, therefore, constructed in order to remove the shortcomings of the former instruments. This instrument, which Yasuaki Asada constructed after the model described in the Ling-t'ai-i-hsiang-chih, was improved by his own device for adjustment and made to meet the requirements for greater handiness and accuracy. This was solely used in observing the time at which celestial bodies passed the meridian. And the transit-instruments Inō used were almost similar in construction to the one Asada had constructed. He had two; one which was fixed at the fundamental station and another which was taken from place to place.

The instrument fixed at the fundamental station was constructed as shown in the diagram. Two square wooden pillars *A* and *B* are planted firmly in the ground in the direction of the meridian. The southern pillar *A* is 3·3 m. in height while the northern pillar *B* is half as high; the distance between the pillars is about equal to the height of the southern pillar. Near the top and about two-thirds of the height from the top on the southern pillar and also near the top on the northern pillar, copper belts *C*, *D*, and *E*

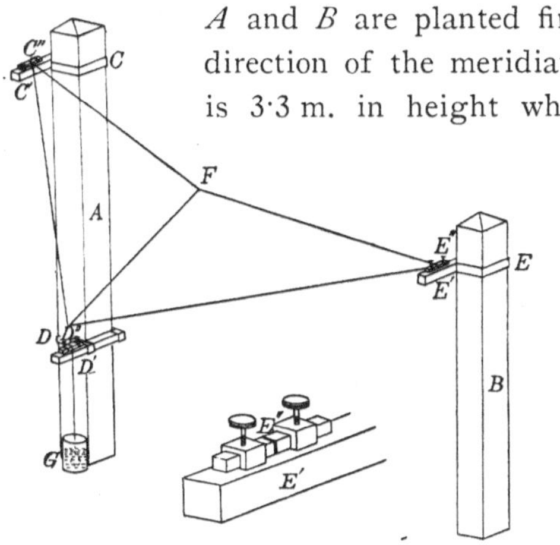

Fig. 25.

are provided; horizontal copper arms C', D', and E' are fixed to the belts, and each of them is mounted by a small copper pillow about 10 cm. in length, about 5 mm. square, and with a narrow groove at the middle. Of the small pillows, C'' on C' is fixed to the horizontal arm while D'' and E'' have a device by which they may be moved by sliding to the right or left and fixed at any desired point on the arms D' and E' respectively. Among the three pillows a fine line is drawn; the line drawn from the small groove on D'' and passing through the groove on E'' and then through that on C'', comes down to D'' where it is connected with the initial end; and the remainder of it is fastened to a plumb dipped into the water-case G provided at the lower part of the southern pillar; besides, a fine line connects F, the central point of the line $C''E''$, and the joint at D'' as shown in the diagram. Such is the principal construction of the instrument.

The essential value of the transit-instrument consists in keeping the plane $C''D''E''F$ formed by the fine line strictly coincident with the meridian plane, and in observing the transit of a celestial body over the meridian taking the plane formed by the fine line as the standard. Therefore, extra care is needed in adjusting the fine line. In order to effect this, another plumb-line is first suspended from C'', and with reference to this plum-line, the pillow D'' is moved right or left until the plane $C''D''E''F$ becomes vertical. Then, taking this provisional vertical triangle formed by the fine line as the standard plane, an observation is made of the upper and lower culminations of a circumpolar star or of any two stars whose difference of right ascensions is known, as already described under the heading of the quadrant; the deviation of the plane of the provisional triangle from the meridian plane is thereby computed; and in accordance with the amount of deviation the position of the small pillow E'' is shifted for correction; with reference to the plane of the revised triangle, a similar observation is repeated and the position of the pillow is adjusted successively until no deviation can be detected.

In observing the transit of the celestial bodies over the meridian, sighting was always done with the naked eye. First, a sofa was placed under the netting; the observer lay on his back upon it, and by shifting his head to the right or left, sighted the netting with one eye until he obtained a position at which the three lines, $C''F$, $D''F$, and $D''E''$, coincided with one another; and with his head steadied there, he waited for the celestial object to come over the line, then measured the instant of its transit. In sighting the transit of fixed stars at night, the lines above were illuminated with a lantern to help the observer. In the case of the transit of the sun a coloured sun-glass was used in observation, and the times of the transit of the west and east limbs and of the apparent centre of the sun were measured, in order to adopt the mean of the results. It is also recorded that sometimes on fine days a piece of light green paper was pasted on a plane-table placed under the fine netting as a means of determining the instant of the sun's transit by the coincidence of the shadows of the three lines, and that this method, however, could not be so accurate as that of sighting with the naked eye, because of the dimness of the shadows.

When observing the transit of celestial bodies one or more assistants were employed who, watching the pendulum-clock, read and recorded the instant of transit as a sign was given by the chief observer. The so-called eye-and-ear method of the present day was never employed.

The portable transit-instrument was constructed in its principal parts exactly like the stationary instrument. The chief difference was that the two pillars were attached to a wooden stand, as shown in Fig. 26, instead of being planted in the ground, and that the length of the southern pillar was 2·1 m. while that

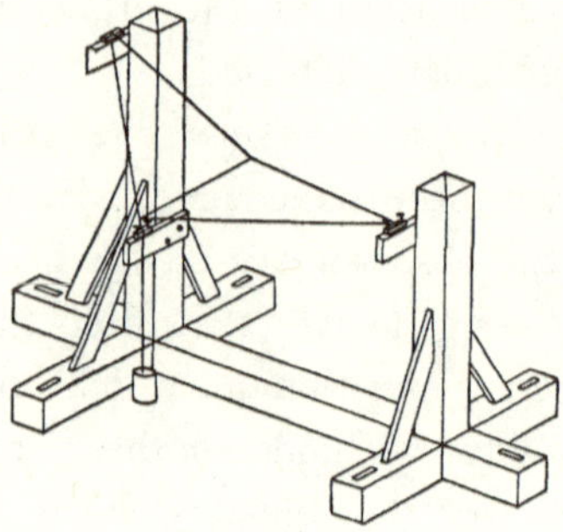

Fig. 26.

of the northern was about 1 m.

Although this portable transit-instrument was so constructed as to be easily taken apart for convenience in transportation, much time and labour* was required to erect it at any desired point and to place the triangular netting strictly in the meridian. This was not used daily like the quadrant, but was applied to practical use only when precise local time was demanded, such as for determining the longitude of a field-station by an observation of the eclipse of the sun, the moon, or the satellites of Jupiter. And since an observation of the transit of a celestial object with the instrument required the greatest skill, Inō himself was in charge of observation except under unavoidable circumstances. And when he had no time for setting up the transit-instrument for an observation of the eclipse of the satellites of Jupiter, it seems, the quadrant was employed instead.

The method by which Inō and others determined local time with this instrument somewhat differed from that employed to-day by which it is computed directly from the observations of the culminations of fixed stars. He directly observed the time at which the sun culminated over the meridian and the ΔT of the pendulum-clock at that instant was determined, taking the local time at that instant to be $0^h\ 0^m\ 0^s$. And the results of observing the culminations of fixed stars for successive nights were made use of in determining the number of oscillations of the pendulum-clock in one day and also in testing the regularity of its motion. In order to eliminate the bad effects caused by the inaccuracy of the right ascensions of fixed stars, observotions were not made at random on different stars each night; but the times of the culminations of some definite stars were observed for several successive nights, when circumstances permitted, and the number of oscillations was usually determined only after comparing the results of observations made on some identical stars for several nights.

The determination of ΔT by the observation of the transit of the sun was got not only on the day on which accurate local time was required, but also on several days before and after. When, owing to bad weather, a satisfactory observation of the transit of the sun was impossible, the ΔT at noon of that day was deduced from the value of the ΔT at the time of the latest observation of the sun combined with the daily number of oscillations of the pendulum-clock determined by the observations of the fixed stars.

As already stated, the adjustment of the transit-instrument under ordinary circumstances was successively performed by means of the results of observations made of the culminations of a pair of fixed stars with a wide difference in declination. For such purpose a mere knowledge of the difference of the right ascensions of the fixed stars was necessary; and if some fundamental observations were made of the relative right ascensions of the more prominent fixed stars numbering a few dozen, it would be no longer necessary to rely upon the antiquated star table given in the I-hsiang-k'ao-ch'êng. In spite of this, Inō who, regarding the declinations given in the I-hsiang-k'ao-ch'êng as untrustworthy, had constructed a table of apparent altitudes based upon his own observations, appears to have made no such attempts upon right ascensions.

(*C*) *The Pendulum-Clock*

The instrument for indicating time which had been used in Japan from time immemorial was the so-called water-clock; and this had been given almost no improvement in a number of centuries. During the last days of the Ashikaga period and the Oda-Toyotomi period, foreign ships brought over a real time-keeper for the first time. This marked a new era in the measurement of time. But the controller of the motion of the wheels, or the most important part of this instrument depended on the oscillation of the horizontal balance; therefore, even the slightest cause considerably affected the oscillation of the horizontal balance and made the motion of the time-piece quite irregular. It was by no means an accurate instrument.

The Ling-t'ai-i-hsiang-chih imported from China, described a method of measuring time by the number of oscillations of a plumb-bob suspended on the end of a fine line. No mechanical device was provided to continue its oscillation; as the amplitude of oscillation decreased, it was increased by thc hand; or a second bob quite similar in dimension was set in motion to succeed the oscillation of the first one, and the numbers of oscillations of the two bobs which swung alternately were added together in keeping time. In spite of such inconvenience, it secured greater accuracy than any previous time-piece, and from about 1760 it became an important instrument to Japanese astronomers in observing the time of eclipses or occultations. No proper method of continuing its oscillations had been invented since then, in spite of great efforts of the students. A pendulum-clock called *suiyōkyūgi* (the swinging-bob instrument) was constructed for the first time about 1789. The inventor of the instrument according to some documents was Yasuaki Asada; and Shigetomi Hazama according to others; opinions are divided on this point. What all the documents agree upon is that the instrument was not introduced from any foreign country, but originated in this country.

A comparison between the extant *suiyōkyūgi* constructed in Japan and the primitive pendulum-clock built in Europe shows that their parts are constructed so exactly similarly that no theory of coincidence may be accepted here. Besides, the ornamental engraving on the index plate can not be Japanese either in its design or in its style. It can not be denied that it is decidedly an imitation of the imported instrument. Even after the declaration of the isolation law by Iyemitsu Tokugawa and the discontinuance of foreign intercourse, not a few curious goods were brought from Europe whenever the privileged Dutch traders came to this country. Especially as Yoshimune Tokugawa had removed the ban against reading Dutch books and considerably alleviated the restriction on importing foreign goods, towards 1770 there were to be found among the common people a number of foreign scientific instruments, such as electric apparatus, microscopes, thermometers, and barometers.

From these facts it can scarcely be believed that the pendulum-clock—an instrument of great importance in daily life and astronomical observations—had alone not been imported into Japan by this time. Perhaps the imported instrument had been secretly treasured by the feudal lords and shut from the knowledge of common people. It may have been that between 1780 and 1790, Asada or some pupil of his happened to see or hear of the instrument, and constructed an imitation.

The pendulum-clock newly built at the Calendaric Office for compiling a new calendar in Kwansei was constructed by Tōzaburō Toda, a mechanician of Kyōto, and the instrument which Inō used was also constructed by the same mechanician about 1796. The latter is still preserved by the Inō family.

The pendulum-clock which Inō used does not differ much in its construction from an ordinary pendulum-clock of to-day except in the arrangement of the escapement, in the manner of suspending the pendulum, in the device for winding up the heavy weight which gives motive power to the instrument, and in the

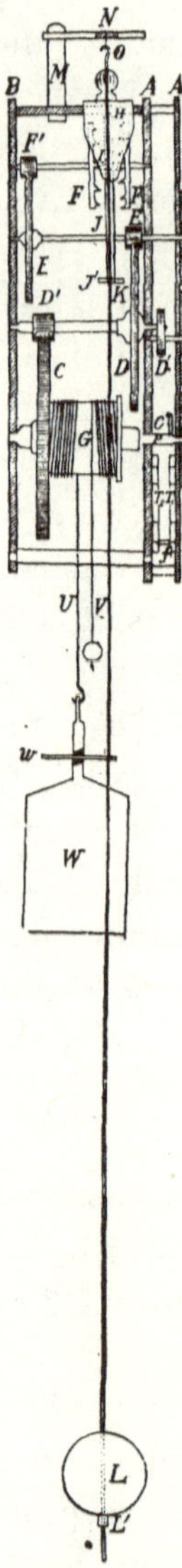

Fig. 27.

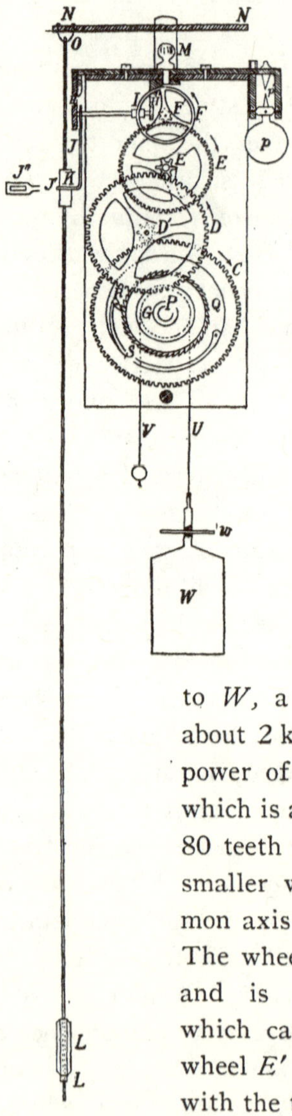

Fig. 28.

dial. The base consists of two brass plates, *A* and *B*, 3mm. thick, 8cm. wide, and 17 cm. long, parallel with each other at a distance of about 5·5 cm., forming a box. Between the two plates, four toothed wheels of brass, *C, D, E,* and *F,* which have parallel horizontal steel axes are installed. The axis of the first wheel *C* is provided with *G,* a drum 3 cm. in diameter, on which a strong silk cord *U* is wound. One end of the cord is fastened to *W,* a cylindric lead weight of about 2 kg. which gives the motive power of the clock. The wheel *C* which is about 8 cm. in diameter has 80 teeth which catch the 8 toothed smaller wheel *D′* fixed on a common axis with the second wheel *D*. The wheel *D* is 6 cm. in diameter, and is provided with 60 teeth which catch the 6 toothed smaller wheel *E′* attached to a common axis with the third wheel *E*. The wheel *E* is about 5 cm. in diameter and has 60 teeth which catch the 6 toothed smaller wheel F' attached to a common axis with the fourth wheel F. Hence the motive power is transmitted in turn from the first wheel to the fourth. The fourth wheel is different from the others in construction: it consists of a pair of rings, facing each other 1 cm. apart; each ring 3 cm. in diameter, has 10 saw-like teeth projecting alternately towards the opposite ring. The catch of the escapement installed between the rings, alternately leaves and catches the teeth in accorddance with the oscillation of the pendulum, so that the wheel F turns by one tooth each time. And each time the wheel turns, some slight motive power is given to the escapement in order to supplement the energy lost through friction and resistance, and in order to maintain the oscillation of the pendulum with a constant amplitude. The escapement consists of a horizontal axis passing between the two rings at right angles to the axis of the wheel F and of a steel piece I with a flat triangular section attached to the former. I is situated between the opposite teeth projecting from the two rings; its delicate adjustment is made by suitably shifting, with the screws, the position of H and H', the brass plates forming the female pivots of the horizontal axis. To one end of the horizontal axis is attached a slender brass stem J; while one end of the stem is fixed to the horizontal axis, the other goes down about 4 cm. where it bends horizontally and has a rectangular hole J'' 7 mm. long and 3 mm. wide, into which the stem of the pendulum is loosely thrust, so that the oscillation of the pendulum may be conducted to the escapement and that the latter may pass the energy it has received from the fourth wheel on to the pendulum.

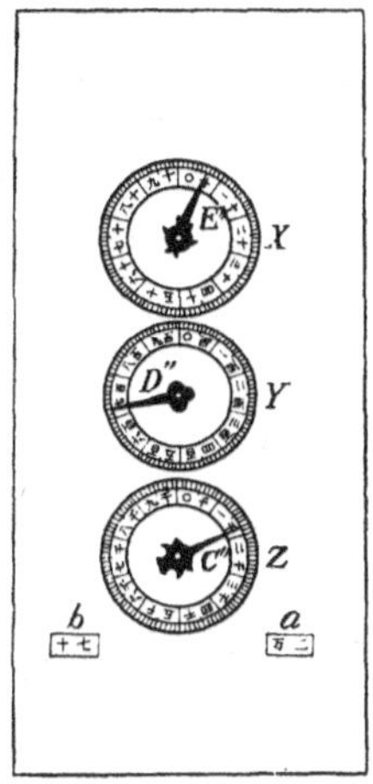

Fig. 29.

The pendulum consists of a stem of iron wire, about 59 cm. in length and about 1·5 mm. in diameter, and of a brass bob about 3·5 cm. in diameter and about 7 mm. in thickness. The position of the bob may be raised or lowered by the female screw L' screwed in the lower end of the wire stem. The upper end is in the form of a hook, and the section K from 7 cm. to 9·5 cm. down from the end is made into a flat form 2·5 cm. in length, 6 mm. in width, and about 1·5 mm. in thickness. The point of suspension of the pendulum is at one end of the horizontal arm NN' projecting from the brass pillar M fixed on the main frame of the instrument, and is situated just above the hole J'' made in the brass stem J. At this point two small holes are bored about 7 mm. apart, and through these holes a fine thread O is strung to form a loop for the purpose. The hook at the upper end of the pendulum is hung on the loop O, and the flat part K of the stem is thrust in the hole J'' so that the pendulum may oscillate within the vertical plane at right angles to the axis of the escapement. During the period in which the pendulum completes an oscillation, each of the two rings of the wheel F turns by one tooth; consequently, during the period the pendulum makes ten complete oscillations, the wheel F completes a rotation; and in turn the wheels E, D, and C, connected with F make a complete rotation during the period the pendulum oscillates 100 times, 1000 times, and 10000 times, respectively.

The drum D on which is wound the silk cord from which hangs the weight W is not directly attached to P, the axis of rotation of the first wheel; they are connected merely by means of the saw-like teeth of the wheel Q fixed on the drum, which catch the brake R on the first wheel, pressed down by the spring-wire S. Therefore, in the direction of rotation caused by the fall of the weight, the first wheel and the drum make the same motion; but when the drum is rotated in the opposite direction for the purpose of winding up the weight, the saw-like teeth pass the brake smoothly and convey no motion to the first wheel. No key is used in winding it up; on the front half of the drum the string *V* is coiled in the direction contrary to that of the silk cord from which hangs the weight, and one end of the string *V* is pulled for the purpose; it is necessary to wind it up once every day.

The indexes indicating the number of oscillations are arranged on *A′*, a brass plate fixed about 6 mm. in front of the main frame *A*, and consist of three dials and two indicators. The hand of the dial *X* for reading figures of the first and second places is attached to the axis of the third wheel *E*; the hand of the dial *Y* for reading figures of the third place is attached to the axis of the auxiliary wheel D_1 connected with the axis of rotation of the second wheel *D*; and the hand of the dial *Z* for reading figures of the fourth place is attached to the axis of the first wheel *C*. Besides, for the purpose of indicating figures of the fifth place and those of the sixth place, there are two revolving plates connected with the axis of the wheel *C*, which are so devised that the figures of the fifth and sixth places marked on them may be seen through the respective small slits. The number of the oscillations of the pendulum indicated by these indexes returns to 0 when 1000000 is reached. Since this pendulum makes about 59500 complete oscillations a day, the number indicated by the indexes recurs in about 17 days. It is no doubt a characteristic of this pendulum-clock that the indexes do not directly indicate the number counted with the unit of 1/10000 day as adopted then or that with the unit of 1/86400 day or second, but give a number counted with quite an arbitrary unit.

The main frame *B* of this instrument is nailed to a stout wooden plate; for practical use the position of the instrument is adjusted by means of *p*, an apparatus for testing verticality, which is attached to the right shoulder of the instrument, and the wooden plate is firmly fixed to the upright pillar. When it is completely set up, the upper part of the instrument, or the section in which the indexes are arranged is covered with a wooden case with glass windows, and the lower part, or the section in which the pendulum and the weight are installed is also covered with a wooden case with a lid, for the purpose of protecting them from wind or dirt.

For recording the time of observation with the pendulum-clock, an assistant was engaged in watching the motion of the indexes of the dials, and with a signal from the chief observer at the instant of observation, he recorded the corresponding number. And the smallest unit of the reading was the time for one complete oscillation of the pendulum, or sometimes that for a half-oscillation. No fraction of these units was ever read.

To convert the number given by the pendulum-clock at any instant to the true solar time, the number indicated on the pendulum-clock at the instant of the sun's transit which was observed most recently, and the number of oscillations of the pendulum during a solar day determined by the transit observations for several days of some fixed stars and the sun, were taken as the basis of calculation, and assuming that the clock made a uniform motion during each day and night, the true solar time for the desired instant was computed by simple proportion.

(*D*) *The Eclipsemeter*

An observation of the four contacts of the solar and the lunar eclipse is effectually made with a common telescope. However, an observation of the magnitude of obscuration or the direction of eclipse at a desired instant during an eclipse requires an instrument with a special arrangement. And the instrument for measuring the magnitude of obscuration of eclipses was called the *sokushoku-teibungi* and that for determining the direction of eclipses the *sokushoku-teihōgi*. These instruments were in the possession of the Astronomical Bureau of Yedo about 1800.

Inō also owned an eclipsemeter similar in construction to the former of those possessed by the Astronomical Bureau, which is still preserved at the Inō home. The eclipsemeter is a sort of micrometer fixed at the focal part of the objective of a telescope. It is entirely of brass, weighing about 1·6 kg., and is composed of 3 parts, *A, B,* and *C,* as shown in the diagram. *A* is a cylinder forming the base of the instrument and serving to fix the instrument to the telescope. *B,* the cardinal part with an apparatus for measurement, consists of a cylindrical hollow case 13·5 cm. in diameter and about 3 cm. thick; by means of the short pipe fixed to a hole about 4·5 cm. in diameter bored at the centre, the case is inserted into *A*; the toothed wheel *Q* attached to *B* and the tangent screw *T* fixed on *A* are arranged to catch each other so that the case *B* may be slowly rotated about the cylinder *A* by the tangent screw *T*. *C* is a cylinder based on the circular plate *R* in the case *B* and projecting out of the case; and is provided with a lens which serves as the eye-piece. Within the case *B,* there is a brass wall *S,* in the centre of which a circular hole about 3 cm. in diameter is made to fasten *W,* the fine silk fibre net strung there in the form of a regular octagon. Besides, the brass wall has two grooves T_1 and T_2 into which the female screws *K* and *L* fastened with the fine needles

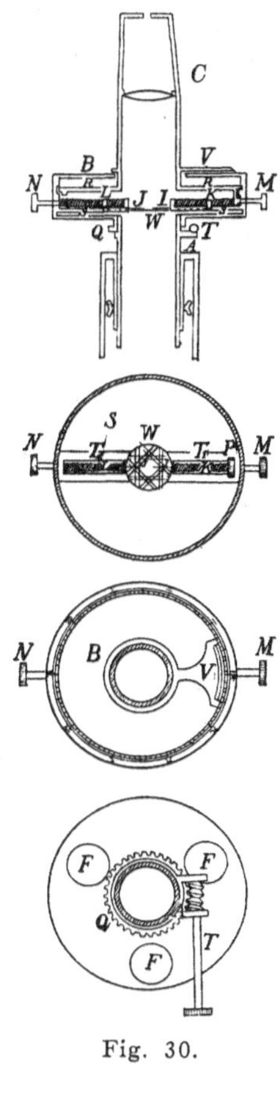

Fig. 30.

I and *J* are thrust respectively; by means of the minute screws *M* and *N* the fine needles *I* and *J* may be moved freely to or from the centre of the octagonal net. A small toothed wheel *P* is attached to the rotation-axis of *M* and its teeth and those provided on the circumference of the plate *R* forming the base of the cylinder *C* catch one another, so that the cylinder is turned with the rotation of *M*; and by referring to the index *V* fastened to the cylinder *C* and the graduation marked on the face of the case *B*, the amount of the rotation of *M*, or the amount of displacement of the fine needle *I* is read. The graduation given on the face of the case *B* divides the circumference into 100 equal parts, and the index *V* is so graduated as to divide 11 parts on the circumference into 10 equal parts, thus forming a vernier. It must be remarked here that a vernier of this form is not used in any other measuring-instrument that Inō employed. On the back of the case *B*, three round windows 2·5 cm. in diameter, covered with mica are bored, so that the fibre net and the fine needles may be illuminated by the lantern from the outside.

The telescope to which this instrument was fixed does not exist to-day. But it was more or less like the one which belonged to the magnetic theodolite, and it was set upon an upright pillar, as shown in Fig. 31, so that it might be turned in any direction.

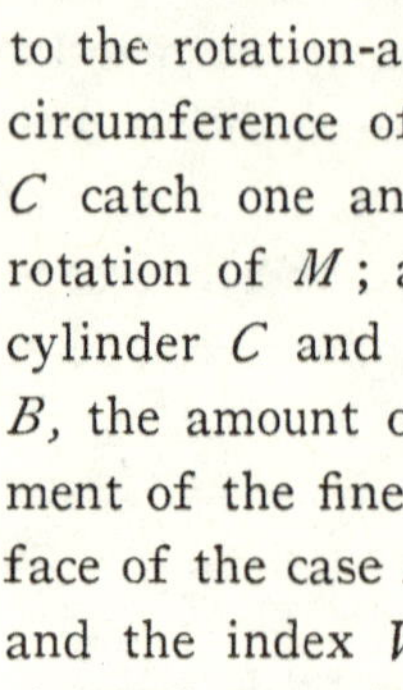

Fig. 31.

In order to measure the magnitude of obscuration at the desired moment during a solar eclipse with the eclipsemeter, the telescope was carefully manipulated to follow the sun before the eclipse and to keep its image precisely in the centre of the octagonal net as shown in Fig. 32; and by turning the screw *N*, the needle *J* was moved up

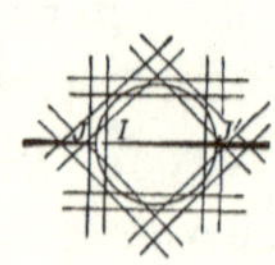

Fig. 32.

till its point touched the outline of the sun's image from the outside; then by turning the screw M, the needle I was moved up till its point touched that of the needle J, and r_1, the indication by the index V for this position was read; then the needle I was moved down until its point touched the outline of the image at the point exactly opposite J and r_0, the indication by the index for this position was then read. The difference between the two readings, or $r_1 - r_0$ was the value corresponding to the diameter of the sun at the time of the observation.

As soon as the phenomenon of an eclipse commenced, not only was the solar image kept precisely at the centre of the octagonal net by properly moving the telescope, but also the case B itself was suitably rotated by means of the tangent screw T until the point of the needle J touched the outline of the solar image from the outside, on the straight line joining the centres of the solar and lunar images; and at the same time the needle I was moved by M until its point touched the outline of the lunar image as shown in Fig. 33.

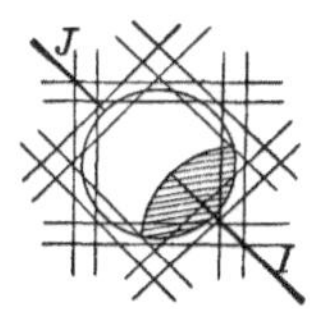

Fig. 33.

At proper intervals, r or the indication for the position of the needle I was read, and each time the assistant measured the corresponding time by the pendulum-clock and computed the magnitudes of obscuration for those instants during the eclipse by

$$\frac{r - r_0}{r_1 - r_0}.$$

Although this eclipsemeter appears at a glance to be a very accurate instrument, it was by no means easy even to manipulate both the horizontal and perpendicular axes of the telescope-stand with the hand so as to follow the solar motion and to keep the solar image strictly in the centre of the octagonal fibre net. It was much more difficult to rotate the eclipsemeter suitably at the same time in order to keep the needle J in touch with the outline of the solar image on the straight line joining the centres of the solar and lunar images. In actual practice, therefore, its accuracy was not a little decreased. Besides, the construction of this instrument involved extraordinary expense. In observing the solar or lunar eclipse at a field-station, Inō employed only one such eclipsemeter, and in addition one or two instruments having much simpler construction which may be called the simpler eclipsemeters were used as controls.

The simpler eclipsemeter, which is presumed to have been invented by Shigetomi Hazama, consisted of a frame which was inserted at the principal focus of the objective of the telescope, and on which several dozen parallel fibres were strung at equal intervals, so that the eclipsed magnitude might be measured by means of these fibres. The construction of one of the few still preserved at the Inō home will be described here. The frame is in the form of a cylindrical box made of a brass plate about 1 mm. thick, 2·5 cm. in inner diameter, and about 1 cm. in length and a brass bottom of the same thickness; at the centre of the bottom a hole about 2 cm. square is made, and along the two opposite sides of the hole, two wires 1·5 mm. in diameter, *AB* and *CD*, with fine screw grooves are fastened parallel to each other. The screw grooves are made by winding an extremely fine copper wire upon a brass wire about 1 mm. in diameter. The screw grooves number about 70 for the space of 1·8 cm.; and fine silk fibres are tightly strung in order between the grooves of the line *AB* and those of the line *CD*. These several dozen lines parallel to one another at equal intervals form a scale for measuring eclipsed magnitudes.

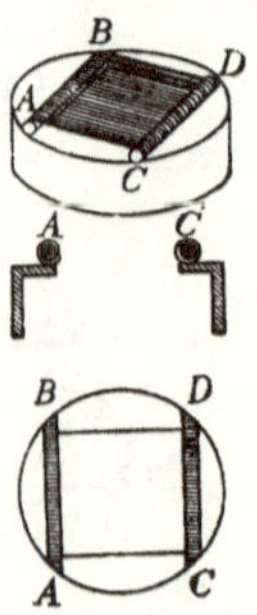

Fig. 34.

In order to use this instrument, the telescope was manipulated for some time before the commencement of an eclipse to follow the movement of the sun and keep its image upon the parallel lines; and the reference numbers w_1 and w_2 of the two parallel lines touching the outline of the image on the opposite outsides were read and the diameter of the sun was determined by w_2-w_1. As soon as an eclipse commenced, the image of the sun was not only kept in touch within w_1 and w_2 by manipulating the telescope, but also by turning the parallel lines suitably, the straight line connecting the centres of the sun and the moon was orientated to be always at right angles to the parallel lines. As the edge of the moon came from w_1 towards w_2 with the progress of the eclipse, each time it came into contact with a parallel line the reference number w of that line was read and a signal was given for the assistant to read and record the corresponding time. The eclipsed magnitude for the moment was computed by $\frac{w-w_1}{w_2-w_1}$.

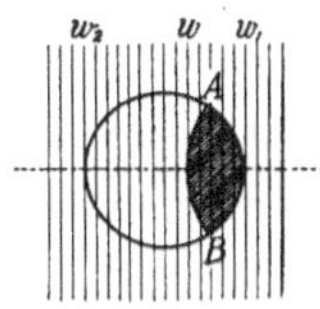

Fig. 35.

When Inō intended to make an observation of a solar or lunar eclipse at a field-station, he stayed there for a few days before and after the day of the eclipse in order to make observations for determining the local time, and on the day of the eclipse all the members of the survey-party were mobilized for the operation. Of all the operations in the observation, the determination of local time was done by Inō himself, the observation of eclipsed magnitudes by a subordinate official and some of his pupils most skilful in the work; and the other members were respectively in charge of the reading of the pendulum-clock, the measurement of the sun's altitude, and other operations.

Now, in the case of a lunar eclipse, the computation of the longitude of the field-station was done simply by taking the mean of the differences between the times for the principal contacts observed at the field-station and those observed at the fundamental station, as shown in the preserved fragmentary record on the subject.

In case of a solar eclipse, it is possible that Inō computed the longitude difference between the field-station and the fundamental station from the observation-materials at two stations according to the method of computing the solar eclipse which Takahashi, his master, had adopted in compiling the Kwansei Calendar. No document is, however, extant giving any practical instance definitely.

(*E*) *The Gnomon*

The gnomon was one of the instruments for measuring the sun's altitude, handed down from the earliest days. Its accuracy had been considerably increased by the great efforts of Yoshitoki Takahashi and Shigetomi Hazama at the compilation of the Kwansei Calendar. The gnomon Inō used was constructed after the instrument newly built for the Astronomical Bureau about 1796. It was permanently set up at his residence at Kuroye-chō in Fukagawa for measuring the meridian altitude of the sun and determining the absolute value of the latitude of this station. Though nothing of this instrument is preserved but the *shaku* scale, its construction may be presumed from documents.

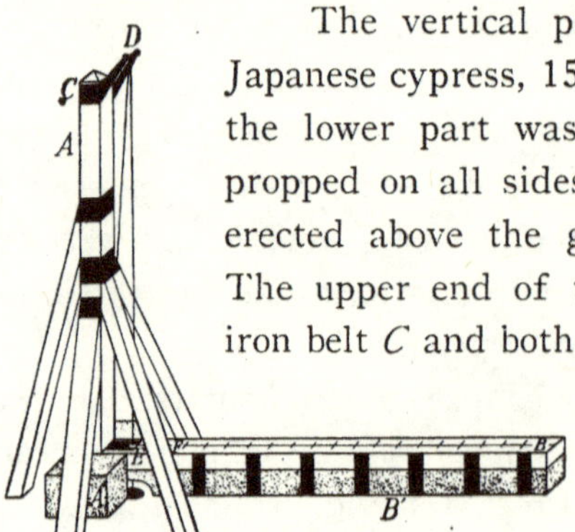

Fig. 36.

The vertical pillar of this instrument was of Japanese cypress, 15 cm. square and about 6 m. long; the lower part was deeply buried in the ground; propped on all sides by five supports, it was firmly erected above the ground to the height of 3·9 m. The upper end of the pillar was wrapped with an iron belt *C* and both ends of the belt extended on the northern side of the pillar, supporting the horizontal bar *D*, a round brass rod about 2 cm. in diameter, which was attached to the ends of the iron belt horizontally in the E.W. direction to intercept the sun light and cast its shadow on the horizontal base. To the centre of the horizontal bar was fastened a fine plumb-line designed to pass through the hole *E* provided on the horizontal base and drop the bob into the water-jar buried in the ground. This was provided for the purpose of testing the verticality of the pillar. The horizontal base consisting of a piece of Japanese cypress about 20 cm. wide, about 12 cm. thick, and about 8·2 m. long was fastened by iron belts to the stone base *B* to the north of the pillar. Along the face a straight line was marked lengthwise, and this was graduated to every 0.5 *shaku* beginning at a point directly below the horizontal bar. This horizontal base when fastened to the main base was so placed as to keep its face strictly horizontal and the lengthwise line *FB* accurately in the direction of the meridian. There was also a long scale for measuring the vertical distance from the horizontal bar to the horizontal base. It consisted of a piece of Japanese cypress 6 cm. wide, 2·5 cm. thick, and about 3·9 m. long; both its ends were covered with brass plates and the upper end was cut into a hook-like form so that it might be hung upon the horizontal bar *D*. The graduation of the scale began at the point corresponding to the axis of the horizontal bar, when it was hung upon the bar, and was made to every 1 *shaku* below it, the last 1 *shaku* at the lower end being subdivided to 1/100 *shaku*. Besides, there were instruments belonging to the gnomon, such as one *shaku* scale, a pair of compasses, a levelling-apparatus for testing the level of the horizontal base, and a small screen for producing the image of the sun more definitely.

The one *shaku* scale has already been described. The instrument for testing the level of the horizontal base consists of a long narrow water-tank, a float called a levelling-boat with an index point, and a levelling-scale—a minutely graduated scale

S fastened vertically to the arm which is erected on the base plate, *T*, as shown in the diagram. For practical use, some sleepers were laid on the horizontal base and upon these the water-tank was set up; then the levelling-scale was placed upon the end of the horizontal base so as to have the scale *S* over the water-tank; the levelling-boat was floated on the water as the index point came opposite *S*, the indication was read; and then the levelling-scale and the levelling-boat were moved 0·5 *shaku*, or 15 cm., at a time on the horizontal base, and the indication by the index point was read each time. In this manner, comparing the readings indicated by the index at various points, the horizontality of the base was examined. If the base was found not to be horizontal, the correction to the normal level was computed for every point of the base.

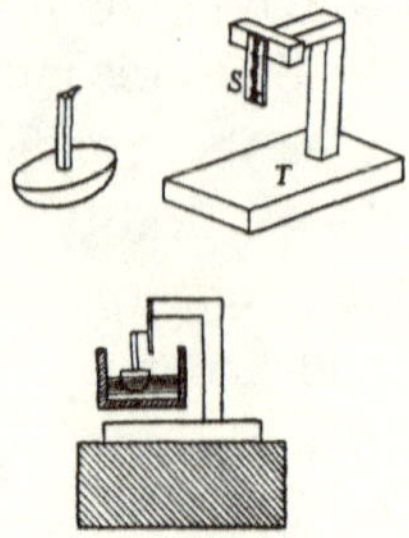

Fig. 37.

The small screen was a brass plate in the form of a round fan having, at its centre, a small hole *H* like a pin-hole, and was attached to a wooden plate by an axis so that the screen might have any inclination. The use of the screen was to project the inverted clear image of the sun and the horizontal bar D upon the surface of the base by means of the hole *H*. The small screen was accompanied by a small wooden box with a face at a slant of about 45° upon which the inverted image of the sun and the horizontal bar were first cast in order to examine whether the image of the bar bisected that of the sun.

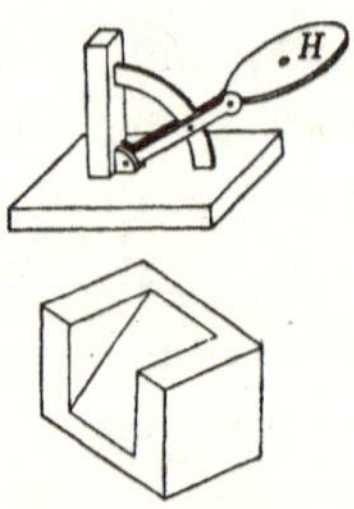

Fig. 38.

Though the gnomon, when set up, was carefully adjusted and all its separate parts were solidly fastened, it was usual to examine it immediately before or after the observation, to see whether the horizontal bar stood at the normal height above the horizontal base, whether the zero mark of the graduation upon the horizontal base was precisely under the horizontal bar, and whether the surface of the horizontal base was absolutely level.

The method of measuring the sun's altitude with this instrument was as follows:—about noon when the shadow of the horizontal bar fell on the meridian, the small screen was placed in the faint shadow of the horizontal bar thrown upon the surface of the horizontal base, and was held in such a slanting as to face directly to the sun so that the hole made in it might project the inverted clear image of the sun and the horizontal bar upon the surface of the horizontal base. And as the small screen was moved slightly up and down in the N.S. direction, the position where the image of the sun was exactly bisected by that of the horizontal bar could be found. This was the position where the centre of the sun, the horizontal bar, and the small hole fell on one straight line, and the image of the bar was clearest. As the surface of the horizontal base was not parallel with the plane of the fan, and did not intersect normally with the sun-beams, the small box with a slanting surface was placed behind the small screen and the images were first projected upon the slanting surface, which was now moved up and down in the N.S. direction jointly with the small screen and adjusted till the image of the horizontal bar projected upon the slanting surface became clearest and exactly bisected the image of the sun; at this point the movement of the small screen was arrested and then the slanting surface removed to let the image of the horizontal bar fall upon the surface of the horizontal base. The point thus obtained was marked as the position of the centre of the sun's image. Though this observation was to be made exactly at noon, it was rapidly repeated several times immediatedly before and after noon in order to decrease observation error, and the position for true noon was determined by averaging the positions marked during each observation. And as this observation demanded quickness and dexterity, it was at that time considered one of the measurements requiring the greatest skill.

When the position of the shadow of the horizontal bar *D*, that is, that of the centre of the sun's image, upon the horizontal base was determined, the apparent altitude of the sun could be easily computed. And in computing the geocentric altitude h_0 from the apparent altitude h, the following formula was adopted:—

$$h_0 = h - r + p \cos h$$

where r was the correction of atmospheric refraction; and p, the horizontal parallax of the sun for which a value $p = 10''$ was adopted.

In order to determine the absolute value for the latitude of his residence at Kuroye-chō, in Fukagawa, Yedo, Inō, on one hand, observed the altitudes of the circumpolar stars both at their upper and lower culminations as already stated; and on the other hand, he computed the geocentric altitude of the sun for each day, from the materials which he obtained by observing with this instrument on the day of the winter solstice and for one or several weeks preceding and following it; they were all reduced to the value corresponding to the altitude at the winter solstice by applying the correction for the change of the declination; and then by a similar operation, he got the geocentric altitude at the summer solstice; and by making use of the altitudes at both solstices, the latitude of the observation station was obtained.

(*F*) *The Telescope*

Besides the telescopes attached to the various special instruments above-mentioned, Inō possessed several independent telescopes, of which two are preserved. One of these consists of four cardboard barrels, as shown in the diagram; it is so constructed as to be shortened when not in use. For the objective a simple double convex lens 4·5 cm. in diameter and with a focal distance of 2·1 m is adopted, but its clear aperture is reduced to 2·6 cm. by a diaphragm. The eye-piece consists of two simple double convex lenses. one 3·5 cm. and the other about 3 cm. in diameter, connected at an interval of about 6·4 cm. and so arranged as to produce an inverted image.

The other one is somewhat smaller than this, the whole length of the barrel being about 1·5 m. and the eyepiece consisting of three simple lenses so arranged as to produce a direct image.

The lenses of these telescopes were made by Zembei Iwahashi, a resident of Kaidzuka in the province of Idzumi, his name being marked on the frames holding the lenses and the barrels. The lenses are all simple ones; the images produced by them, however, are comparatively fine and clear in spite of the presence of the chromatic aberration.

Fig. 39.

These telescopes were used in observing what required no special apparatus within the telescopes, such as the times of the four contacts of a solar or lunar eclipse, or the times of the eclipse or occultation of Jupiter's satellites.

6 The Result of His Astronomical Observations

The observations Inō carried out with the above-mentioned instruments during the latter half of his life were varied and numerous. Above all, the results of his determination of longitude and latitude directly connected with the survey-work of Japan may be summarized as follows:—

(*A*) *His Latitude Measurements*

The fundamental station is the only one for which Inō made an absolute measurement of latitude. And the result has already been stated elsewhere; the value obtained was

Kuroye-chō, Fukagawa, Yedo ....35° 40′ 30″*N*.

Besides this, Inō's diary shows that, for about a month and a half at the beginning of 1806, he sojourned at Shimo-no-chō, Okayama, in the province of Bizen and was engaged in astronomical observation day and night; and it may be assumed that he observed the upper and lower culminations of Polaris for obtaining the absolute latitude of this station. Unfortunately, there exists no document to confirm this.

The latitudes for the other stations were all determined by a relative measurement based on the latitude of the fundamental station. The number of the stations given in his work, "Enkai-jissoku-roku" (The Record of the Coast-Survey) alone is no less than 1200. As this record omits the latitudes for the stations at which a sufficient number of fixed stars were not observed or the observation conditions were not satisfactory to make the results thoroughly reliable, the total number of the stations at which he measured latitude must be much greater than this. If the number of the stations at which his all night observation resulted in utter failure due to bad weather were included, it would be still greater. The stations given in the record are distributed among the districts as follows:—

the Kinai	district................	about	40	stations
,, Tōkai	,, ...	,,	180	,,
,, Tōsan	,, ...	,,	180	,,
,, Hokuriku	,, ...	,,	80	,,
,, San-in	,, ...	,,	80	,,

the San-yō district	....	about 150	stations
,, Nankai ,,	....	,, 160	,,
,, Saikai ,,	....	,, 320	,,
,, Yezo ,,	....	,, 30	,,

This shows that, notwithstanding his utmost efforts throughout, his stations for latitude measurement were fewer in the eastern half of the country and of a greater number in the western half, because he was not able to spend more days in measuring a district during his earlier years of land-surveying. And of all these stations, numbering more than 1200, there were about twenty at which he made observation for several nights to obtain the most accurate results for determining the length of one degree of the meridian arc, for correcting geodetic error, or for some other reasons which he considered especially important. These are given below:

Station	°	′	″
Mimmaya, Tsugaru County, Mutsu Province	41°	11′	50″
Yorodzuchō, Noshiro Port, Yamamoto County, Dewa Prov. ..	40	12	30
Gofukuchō, Takata, Kubiki County, Echigo Prov.	37	7	0
Aramachi, Tokoroguchi, Kashima County, Noto Prov.	37	2	30
Nishihamachō, Tsuruga, Tsuruga County, Echizen Prov. ..	35	39	30
Shinsen-enchō, Kyōto, Yamashiro Prov.	35	0	36
Yajitachō, Shimoda, Kamo County, Idzu Prov.	34	40	30
Shimo-no-chō, Okayama, Mino Country, Bizen Prov.	34	40	30
Tondayachō, Ōsaka, Settsu Prov.	34	40	0
Fujinogō, Toba, Tōshi County, Shima Prov.	34	28	30
Sakaichō Shichōme, Hiroshima, Numada County, Aki Prov.	34	24	0
Nishikiura, Muro County, Kii Prov.	34	13	30
Nambuchō, Akamagaseki, Toyoura County, Nagato Prov. ..	33	57	30
Sendōmachi, Kokura, Kiku County, Buzen Prov.	33	53	30
Hon-machi, Tanabe, Muro County, Kii Prov.	33	44	0
Tanezakimachi, Kōchi, Tosa County, Tosa Prov.	33	34	0
Hon-machi Nichōme, Uwajima, Uwa County, Iyo Prov.	33	14	0
Hon-machi Shichōme, Ōmura, Sonoki County, Hizen Prov.	32	54	30
Rokasuchō, Nagasaki, Sonoki County, Hizen Prov.	32	45	0
Kurumachō, Kagoshima, Kagoshima County, Satsuma Prov.	31	36	6

Among the stations considered the most important, besides Kyōto and Ōsaka which were sub-fundamental stations, were Mimmaya and Noshiro in the north-eastern district, and Nagasaki and Kagoshima in the south-western.

(*B*) *His Longitude Measurements*

The instances in which Inō succeeded in computing longitude by means of solar and lunar eclipses and other astronomical phenomena were so extremely rare that they are not to be mentioned in the same breath with his latitude measurements in their number or results. It is true that the solar and lunar eclipses that might have been observed in Japan between 1800 and 1816 numbered more than twenty. There occurred, while the survey-party was in field-work, four solar eclipses and nine lunar eclipses—thirteen in all. Again, those which were observed both at a field-station and the fundamental station in Yedo or at a sub-fundamental station in Ōsaka simultaneously numbered only two solar eclipses and three lunar eclipses—five in all. Of these five, the one observed at Nishikiura in Kii Province on July 12th, 1805, and another observed at Matsuida in Kōdzuke Province on Oct. 23rd, 1809, were both partial lunar eclipses which happened in the horizon; as this phenomenon naturally difficult to observe was observed under especially unfavourable conditions on these two occasions, it is evident that the longitudes thus derived did not serve as materials for correcting geodetic error. The total lunar eclipse which the survey-party not including Inō himself observed on Dec. 16th, 1815, at Shimoda in Idzu Province under favourable conditions, was also well observed in Yedo and Ōsaka; but an unusual error in observation defeated the purpose of utilizing the phenomenon. The solar eclipse of Nov. 18th, 1808, was observed fairly successfully both at Tomariura in Sanuki Province and at Ōsaka; however, the distance between the two stations was not great enough to make it necessary to determine the difference of their longitudes astronomically. And the solar eclipse of April 4th, 1810, observed at Hatoura in Buzen Province was incomplete, since unfavourable weather obstructed the measurement of the first stage of the eclipse, and only some phases of the obscuration were observed now and then between clouds. It is very doubtful, therefore, whether any reliable result was obtained.

After all, it seems that Inō succeeded only once or twice in obtaining inaccurate longitude differences by the results of the observation of a solar or lunar eclipse, and that he was not once able to make use of an eclipse in correcting for geodetic error as the result was not sufficiently accurate for the purpose.

As for the results of the observation of the eclipses or occultations of Jupiter's satellites newly commenced after 1805, it was not often that he fulfilled his purpose when surveying the Kyūshū district, although he stayed several days at one station to observe such phenomena and did his very best to do so. At any rate, observations were performed at 15 or 16 stations, according to his journals. Even though some omissions in his journals may be assumed, the number would not be doubled. Of all these observations, if those carried out under unfavourable conditions or resulting in inaccuracy due to some obstacles, were omitted, the cases in which fairly successful observations were made would not be quite half as many. Still smaller would be the number of the cases in which a simultaneous observation was successfully made at the fundamental station in Yedo—perhaps only five or six at the most.

And besides, since no observation record of the satellites of Jupiter or calculation-note of longitudes has been preserved unfortunately, it is impossible to know definitely what results Inō attributed to such and such stations. But, in general, an accurate observation of the eclipses or occultations of Jupiter's satellites involved extreme difficulties, and observation error even under the most favourable condition would be too great to serve the purpose of correcting his geodetic errors. It goes without saying, therefore, that in spite of his possible success in obtaining the longitudes of several stations based on his observation-materials of Jupiter's satellites as a result of his tremendous efforts, the accuracy of the longitudes was not such as to satisfy his original purpose of controlling the geodetic error and of supervising the maps. As will be noted later, the map of the whole country failed to be projected in a proper way; and in spite of the gross errors introduced on the map in the longitudes of the stations in the north-eastern and south-western districts, no attention was paid by the experts who carelessly overlooked them. This proves that there existed no station where longitude was accurately determined astronomically and that, at the same time, all attempts to determine longitude were unsuccessful.

Although Inō's attempts to determine the longitudes of his field-stations by his own observation-materials obtained there proved utterly fruitless, he succeeded in computing the longitude difference between Yedo and Ōsaka fairly accurately, by the observation-materials of solar and lunar eclipses obtained during a number of years by Asada and other masters; and this, it seems, proved helpful in testing the accuracy of his geodetic results.

The longitude difference between Yedo and Ōsaka had been calculated temporarily by means of the solar and lunar eclipses and other astronomical phenomena after the compilation of the Jōkyō Calendar in 1684, but a thoroughly reliable result had not been obtained before the end of the 18th century. Even in the calendar compiled in the Kwansei era, the longitude difference between the two stations had been given only roughly. After this, however, important observation-materials in determining the longitude difference had gradually increased. During the period between 1795 when the new astronomical instruments were adopted at the Astronomical Bureau in Yedo, and 1818, the year in which Inō died, the number of solar eclipses simultaneously observed in Yedo and Ōsaka was 7, while that of lunar eclipses was as high as 11 or 12. These observations in Yedo were all made properly at the Astronomical Bureau, and some of them were also carried out voluntarily at his own residence by Inō, while nearly all of those in Ōsaka were made by Shigetomi Hazama or Shigeyoshi (Jūshin), his son, at their own residence at Tondayachō in Nagahori. Independent observations down to about 1808 made by Ryūtatsu Asada, adopted son of Gōryū Asada, and Nobuaki (Shintō) Adachi were also available at that time, as first-rate materials, together with those obtained by the Hazamas. And by means of these observation-materials, Inō, in his last years, adopted 4° 19′ as the value of the longitude difference between the Yedo Astronomical Bureau and Tondayachō in Ōsaka, and 4° 4′ as the value of that between the Yedo Astronomical Bureau and the site of the Calendaric Office at Nishi-Sanjō-dai in Kyōto. It should be remarked that these values almost coincide with those given in the map of the whole country—namely, a little over 4° 18′ for the longitude difference between Yedo and Ōsaka, and a little over 4° 3′ for that between Yedo and Kyōto.

CHAPTER V

HIS ADJUSTMENT OF SURVEY ERRORS AND HIS METHOD OF MAP-CONSTRUCTION

1 Determination of the Length of One Degree of the Meridian Arc

Inō's success in surveying the whole of Japan without introducing gross errors may be chiefly attributed to the pains he took to prevent an accumulation of geodetic errors by the astronomical determination of latitude. If the adopted length of a degree of the meridian arc had been erroneous, it would not have proved effective in correcting the geodetic errors, but would have aggravated confusion in the adjustment. It may be asserted, therefore, that his success in land-surveying mostly depended upon his accurate determination of the length of one degree of the meridian arc.

The theory that the form of the surface of the earth is not a perfect sphere had been introduced from Europe by the time Inō came up to Yedo. It had been known especially since 1803 when Lalande's Astronomia was brought into the Astronomical Bureau, that the earth's surface is in the form of an oblate ellipsoid of revolution; this new theory was approved by Inō as well as by his great predecessors. For all this, Inō consistently assumed in his determination of the length of the meridian arc and other computations concerning latitude and longitude that the surface of the earth forms a sphere. And for the purpose of determining one degree length, he never conducted a special survey, but he merely chose what suitably served his purpose out of the survey-materials of the coast and the highways. The method of calculation was very simple,—namely, according to the method described under the heading of his field-note reductions, he computed the N. S. component length of a long section of a surveyed route most suited for determining one degree length; this value he divided by the difference of the latitudes astronomically measured at both end points; thus computing the value of one degree length respectively for various sections and then properly combining all the results, he determined the final value.

It was in the year 1800 indeed that Inō for the first time computed the length of one degree of the meridian arc; availing himself of the survey-materials of the south-eastern coast of Yezo Island and the Ōshū Highway which he had roughly surveyed in the same year, he obtained about 27 *ri* as a result. Since this was simply an approximate value calculated on the results of a rough survey, Inō repeated his computations on the survey-materials of the eastern coast of the Main Island and the Ōshū Highway which he surveyed in the following year, 1801, till he obtained 28·2 *ri* for the value.

Among the materials Inō made use of in his computations of one degree length that year, the route between Fukagawa in Yedo and Noheji by way of Utsunomiya, Shirakawa, Sendai, and Morioka, ran over 5 degrees almost in the meridional direction, and the road was comparatively smooth and straight; hence it was the very best route in determining one degree length of the meridian arc. But as the distance between Fukagawa and Sendai was surveyed that year chiefly with the perambulator and that to the north of Sendai was mostly determined by pacing, the N. S. component between the two extreme points was necessarily quite inaccurate; therefore, the length of one degree of the meridian arc obtained from this route should not be regarded as accurate. Besides, though the distance along the eastern coast-line of Mutsu was thoroughly surveyed with the measuring-rope, great accuracy could not be expected on account of the precipitousness which extremely embarrassed the surveyors. The surveyed line along the coast of Idzu was of the same kind. The line along the coast of Hidachi and Shimōsa was rather favourable in topographical features and surveying conditions for the purpose; but the distance being too small, an accurate result could not be obtained from it.

For this reason, 28·2 *ri,* the value Inō determined after combining the results of his surveying in 1801, was not to be readily credited. Even Yoshitoki Takahashi had his doubts as to this evaluation.

In spite of his strong belief that the value he had obtained involved no serious error, Inō also considered Takahashi's doubts reasonable. In his survey of Dewa and Echigo Provinces in 1802, he desired to confirm the value by the new materials. In his measurement of distance and latitude this time, he took every care that his measurement should admit of no doubt whatever; and from the new materials he computed the length of one degree. All records of the computations, except a summary for the section between Fukagawa in Yedo and Shirakawa by way of Utsunomiya, have been lost. According to this summary, he obtained 40·6111 *ri* for the N. S. component length of that section and 1° 26′ 30″ for the latitude difference, from which he got 28·17 *ri* for the length of one degree. Though no minute record of the respective results obtained from his survey-materials of the other sections he obtained in the same year has been found, the explanatory notes on the map drawn in 1804 suggest that the value obtained from every section was 28·2 *ri* or very close to it. Inō thus confirmed the accurateness of the value he had computed the year before; and after confirming it again by a similar computation on the survey-materials obtained in the following year, or 1803, he finally put 28·2 *ri* as the set value for the length of one degree of the meridian arc.

Whether this value 28·2 *ri* means exactly 28.20 *ri* as a precise value obtained by properly combining the results of repeated computations on reliable survey-materials, or a value merely close enough to be represented by 28·2 *ri*, is not quite clear since no record of computations has been preserved. According to some extant documents concerning his computations of one degree length, Inō made it a point to calculate down to the second or third place below *ri*, which fact proves that the figure of the second or third place below *ri* which Inō ignored in taking 28·2 *ri* as the set value was so small that he was quite willing to dispose of it. That is, the value Inō finally got must have been very close to 28·20 *ri*.

As to the determination of one degree length Yoshitoki Takahashi took an extremely discreet attitude; when Inō, by the survey-materials he secured with utmost care in 1802, proved the correctness of the value he had obtained the year before, even then, Takahashi hesitated to accept it as the set value. It was by mere chance that he came across Lalande's Astronomia which informed him of the value lately obtained in Europe. Seeing that it coincided with the value Inō had reached, Takahashi now acknowledged Inō's great ability and accepted this evaluation as the invariable length of one degree of the meridian arc.

2 The Instruments and Paper Used in His Map-Making

In constructing a map by combining astronomical and geodetic materials, Inō by no means determined the final positions of the various points by adjusting their co-ordinates exclusively by means of computation; while on one hand, computing the

the survey-materials to some extent, he, on the other hand, drew drafts one after another from the materials, and made it a point to examine the drafts for discrepancies, to investigate their cause if any were found, to make suitable corrections to the drafts, to make the corresponding corrections to the results of computations and then to proceed with further calculations. So it was necessary to draw a draft minutely and accurately and to take the utmost care in handling the paper so as to minimize expansion and contraction. The instruments employed in map-making and the pains taken concerning drawing-paper will be discussed below.

The Drawing-Board. No drawing-board or record as to its construction has been preserved. From fragmentary documents, however, it may be assumed that the size of Inō's drawing-boards varied according to their respective demands and the one used for drawing a large map was 1·5 m. in width, 3 m. in length, and of very solid construction.

The Rulers. Well-seasoned Japanese cypress with fine straight grain or well-seasoned oak with straight fibre was adopted for making rulers because of their slight danger of warping. Two pieces of such wood were whetted against each other until their edges were perfectly straight. The length was not uniform, but generally about 1 m., and their width and thickness differed accordingly. Besides these rulers, Inō seems to have used rectangular rulers and parallel rulers also, though neither they nor any records of them are extant to-day.

The Drawing-Scales. The scales used in mapping have all been lost sight of. Their shape, however, may be guessed from the description and diagram in the "Ryōchi-denshūroku." They were constructed as follows: a brass plate about 2 cm. wide and about 1·5 mm. thick was scraped along one edge until it became about 0·5 mm. in thickness; this thin edge was graduated to every 2 *rin* or 2/1000 *shaku*; the length was 0·5 *shaku* or 1 *shaku* according to the demands. The scale strictly coincided with that employed as the standard scale in practical surveying, and it was specially made by a skilful mechanician that it might be graduated with perfect regularity.

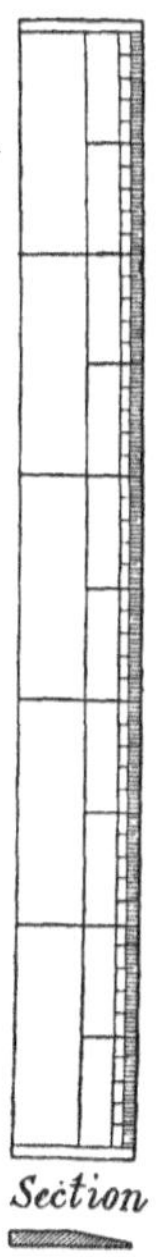

Fig. 40.

The Reducing-Scales. These were used in reducing a draft to 1/6 or 1/12, that is, in reducing a draft drawn in the reduced scale of the large maps, to that of the middle-sized or small maps. We here describe the scale used in reducing a draft to 1/6. A plate about 3 *shaku* in length with a section similar to that of a normal scale was graduated along one edge to every 2/1000 *shaku* from 0 to 0·5 *shaku*, but the remainder was graduated at a rate six times as great as the normal graduation to every 5/10000 fictitious *shaku* (a fictitious *shaku*=6 *shaku*) or 3/1000 *shaku* of the normal scale for the range extending to 0·5 fictitious *shaku* or to the end *E* corresponding to 3 *shaku* by the normal scale (Fig. 41); the scale graduated in fictitious *shaku* or at the rate of six times the normal scale was called a 1/6 reducing-scale: thus the length from 0 to a given point of graduation on the 1/6 reducing-scale was six times as great as the length from 0 to a similarly numbered point of graduation on the normal scale.

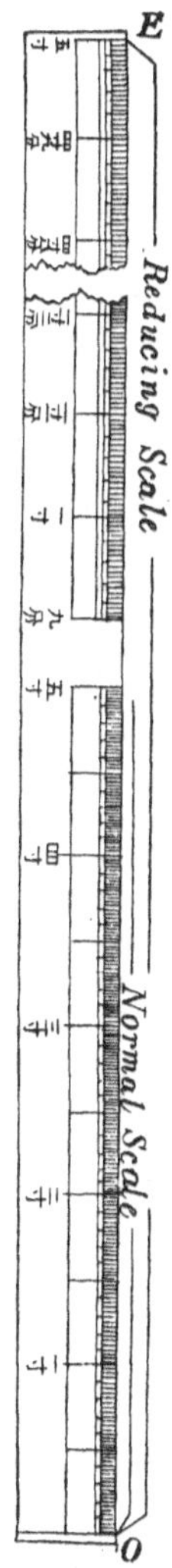

Fig. 41.

The method of using a reducing-scale was extremely simple. In reducing the map **A** to a sixth, for instance, the reducing-paper **B** was pasted at a convenient corner of the map; at *C* a point on **B** convenient for drawing, a small hole was bored with a needle-point as the centre for reducing the map. The reducing-scale to a sixth was then so placed that one end 0 coincided with C and the distance from this point to the point P on the map **A** was measured by the fictitious graduation of the reducing-scale; then the corresponding distance designated by the same figures was measured by the normal scale graduated near the end 0 of the scale; and at that point a pin-hole p was made. In a similar manner, q, r, s, and t, or the points corresponding to Q, R, S, and T on the map **A**, were marked on the map **B**, and by connecting these points, a reduced reproduction of the line $PQRST$ on the map **A** was obtained. Because this method involved no calculation whatever and was free from an accumulation of drawing errors, it was regarded as the simplest and most accurate of all the reducing-methods then adopted in Japan. It is to be remarked here that no reducing-instrument such as a pantograph was then known in Japan.

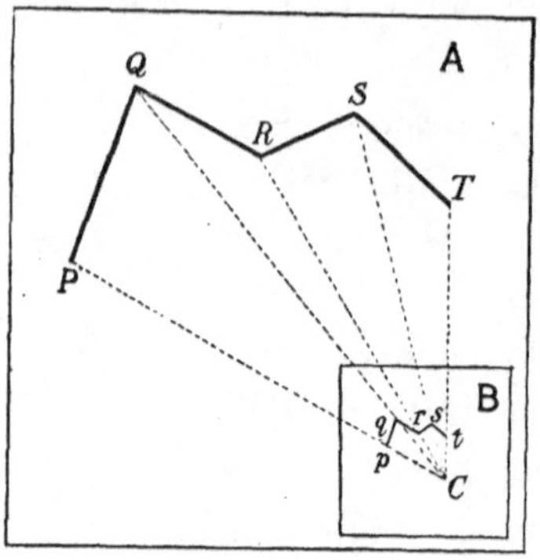

Fig. 42.

A reducing-scale, if used reversely, may be utilized in drawing a magnified map. But since this was apt to introduce considerable drawing error in the magnified map, Inō never availed himself of the reversed use of a reducing-scale.

The Protractors. Though the protractors Inō used were by no means all of the same shape, the usual one consisted of a thin semi-circular plate graduated along the arc, almost similar in construction to the one adopted nowadays. The material of the plate was varied, it being sometimes of cardboard, brass, or wood. In drawing, however, a brass protractor proved inconvenient because of its heaviness, and mostly cardboard protractors were used; not a few of them being preserved. Their usual size is from 4·5 cm. to 6 cm. in radius, the largest being 10 cm. in radius; some wooden ones are still larger. They are graduated to every degree, but with protractors of greater radii, each degree is subdivided into halves or thirds. At the centre of the graduated semi-circle, a square hole about 1 cm. on each side is cut; two fine hairs are fixed on it to form a cross, as shown in the diagram, *C* at the intersection strictly coinciding with the centre of the graduated semi-circle, because this facilitates, in practical use, putting the centre of the protractor at a given point. Moreover, the straight edge *AB* is made to be in strict parallel with the radial line showing the meridian of the protractor, so that, in mapping, the line *AB* is always kept parallel with the meridian on the paper for the purpose of setting up the protractor in the right direction on the map.

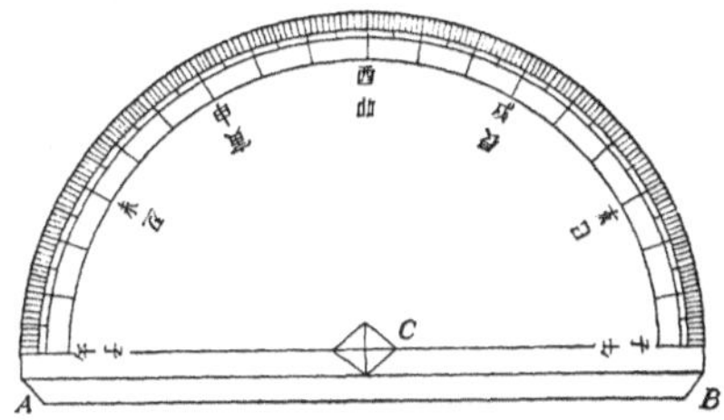

Fig. 43.

The Compasses. The introduction of this instrument into our country dates from a very early time as is shown by the fact that as early as about two centuries ago compasses were used by land-surveyors under the European name *kompatsu* not only for measuring and transferring any length in drawing, but also for measuring distances; so much so that a new term, the *kompatsu* method, was soon coined. Inō employed these only in drawing, and never in any other way described in the so-called *kompatsu* method. An examination of the fragments of the instruments preserved at the Inō home shows that they had been manufactured no doubt by Japanese mechanicians. Their construction, however, is thoroughly of foreign style, not differing much from the compasses we use to-day in drawing.

The Drawing-Compasses, Drawing-Pens, and Dotted-Line Drawers. Like the compasses, these instruments were all of western style, being almost similar in construction and shape, as may be inferred from the fragments preserved at the Inō home.

As for other materials for drawing, some pencils were imported from foreign countries at that time, and there is some indication that an attempt was made to imitate them in this country. We find some pieces of coarse graphite in the case which contains the fragments of Inō's stationery. However, there is no indication of his having used a pencil in practical drawing. An examination of the drafts of his maps still extant shows that the lines were first marked on the paper with such an instrument as a drawing-pen without ink, and only in finishing, black Indian ink or vermilion was applied on the necessary parts.

The Drawing-Paper. Since Japanese paper is apt to expand or contract considerably according to the state of humidity, Inō took the greatest pains in his operations, in order to minimize its expansion and contraction when drawing maps—especially those forming fundamental materials such as drafts and the so-called "combined" maps. An examination of the extant maps and a reference to the "Ryōchi-denshūroku" show that a paper called *nishinouchi* was used for drawing drafts and "combined" maps; sheets of this paper were first pieced together to make a suitable size in accordance with the needs of the case, it was then damped with water and spread on a board and left in that state for about five days so that it might freely expand or contract, and then only was a map drawn on it. Such an original map was never lined. When an original map was completed and a fair copy of it was demanded, either a paper called *minogami* to which a solution of alum had been applied or *tōshi* (Chinese paper), which had been moistened with water and spread on a board, was used; the original map was placed upon such paper and, by pricking the map with a needle-point, it was transferred to the paper. Some such copies were lined and others were not. The original copies presented to the Shogunate Government and their duplicates preserved at the Inō home were all substantially lined with such paper as *tōshi*, *kizuki*, or *maniai*. Such lining was not attended to on the completion of the maps; but after the paper had been lined and sufficiently exposed to its natural expansion and contraction, a map was drawn upon it.

3 His Method of Drawing Maps of Small Sections

For drawing a small-section map, on a suitably sized sheet of *nishinouchi* paper which had been moistened with water and dried, a number of straight lines were drawn, in white, strictly parallel with one another and approximately 6 cm. apart, which represented either the E. W. direction or the N. S. direction as suited his convenience—the standard lines in case of using a protractor on the paper—and at right angles to which, straight lines could be drawn representing the N. S. direction or the E. W. direction. In projecting a small-section map the surface of the earth was regarded as a perfect plane.

The area of land to be projected on a sheet of paper was a section the survey-party commonly covered in one day. First, on the suitable position of the paper was determined A a point corresponding to (A) an end of the surveyed line in the section; along the E. W. line and the N. S. line from this point, the lengths AC and CB, equal to the E. W. component length and the N. S. component length (the values computed from the field-notes and reduced to the scale of 1:36000), with which (B) the other end of the surveyed line in the section parted from (A), were measured and the position B corresponding to (B) was determined. Then A was taken as the centre of the protractor, its orientation being adjusted by referring to a white line; then, according to the contents of the field-notes, a white line Ap was drawn with the azimuth of the first segment of the surveyed line starting from (A); on this line, AP equal to the length of the segment (after being corrected for inclination and reduced to the scale of 1:36000) was obtained and the position of P was determined. Then the centre of the protractor was moved to P and its orientation adjusted; and according to the contents of the field-notes, a similar method was adopted to determine the line PQ corresponding to the second segment of the surveyed line starting from the point P. Repeating the same operation, all the segments of the surveyed line were reproduced on the paper until B or the final point of this section was reached. Then the position of B on the paper thus obtained should always coincide with the point B which had been determined beforehand on the paper by means of the E. W. component length and the N. S. component length (B) had as against (A). If they failed to coincide, it indicated some mistake in drawing or computation of the E. W. or N. S. component length. Therefore, either was examined for the mistake and was corrected until the right position of B was secured.

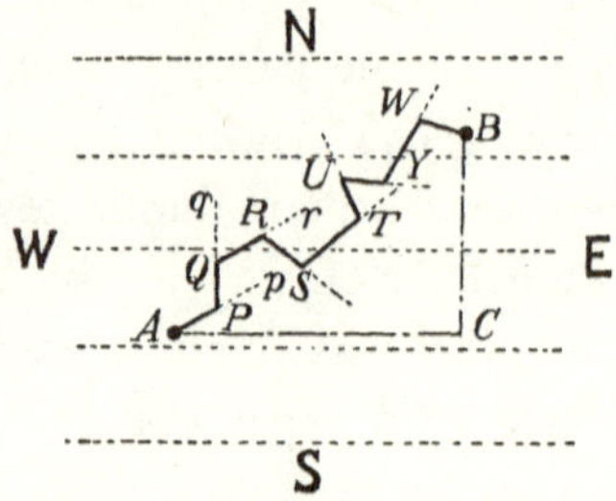

Fig. 44.

After all the segments of the surveyed line between (A) and (B) had been drawn and absence of mistakes in drawing or computation confirmed, a white line was drawn, in each direction of observed azimuths of the mountains, trees, temples, and shrines in the neighbourhood taken from various points on the surveyed line. Unless there were some error in the observation of the azimuths of these objects or in the measurement of the surveyed line itself, the azimuth lines drawn from different points on the surveyed line towards the same object should meet at one point.

In practice, however, some slight error in surveying was unavoidable and it was seldom that the azimuth lines towards the same object exactly converged on one point; generally there was some divergence. According to the aspects of the divergence the most probable point of intersection was judged, or a suitable correction to the position of the surveyed line was made. An examination of the numerous drafts now kept at the Inō home shows that the divergence of the intersection of the azimuth lines was generally small enough to be ignored.

Despite Inō's efforts in every way to eliminate mistakes in taking readings or in recording them, some were of course sometimes introduced. The materials of the azimuth observation made on the objects in the vicinity proved exceedingly helpful in detecting and correcting mistakes committed unconsciously. For instance, when a careless mistake was committed at a certain point of the surveyed line, the converging point of the azimuth lines towards a certain object observed from the various points up to the point at which the mistake was committed, and the converging point of the azimuth lines towards the same object observed from the various points, lying after the point at which the mistake lay, would be found to be quite different. Therefore, the introduction of the mistake would be readily detected and a suitable correction would be made. As shown in the diagram, if the observed azimuth lines towards one object from P, Q, and R on the surveyed line between A and B centred on the point E and those

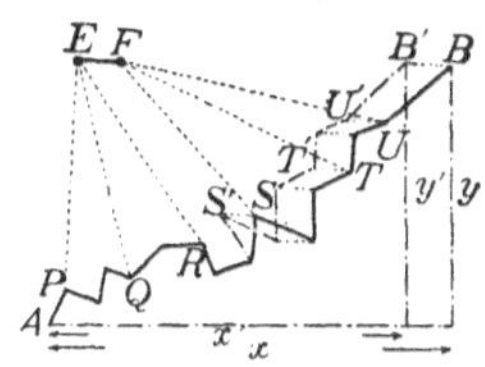

Fig. 45.

towards the same object from S, T, and U centred on F or a point different from E, it would definitely indicate some mistake either in surveying or recording, committed between R and S on the surveyed line. In order to adjust this error, E and F were connected, and straight lines parallel to EF were drawn through the various points after S; each such point was shifted along the line as much as the length of EF and finally the corrected positions S', T', U', and B' were determined. When the distance between R and S consisted not of a single segment of the surveyed line, but of a combination of more than one segment as shown in the diagram, it was impossible to tell on which segment the mistake had been committed; thus, it was impossible definitely to correct the segments of the surveyed line between R and S. But the inaccuracy arising from this meant an inaccuracy concerning only that part, and the remaining part or the points from S to B had to be exactly corrected. Some of the drafts preserved at the Inō home show marks of correction by such a method. If Inō made any correction, he did so deliberately only after he had given it sufficient consideration and had become convinced of justification in doing so.

After examining for errors and mistakes and correcting these, he measured on the map the E. W. component length x' and the N. S. component length y' with which B' one end of the corrected surveyed line deviated from the other end A. Thus, he obtained the revised E. W. and N. S. component lengths, the numerical values of which he wrote on the map lest errors should arise in copying, due to the contraction of the paper. Such revised small-section maps served as the originals for making a *yosedzu* or "combined" map. This method of projection and correction Inō applied to the map of a small section lying in an area about 10 km. square was most effective. It stands unrivalled when compared with the other methods adopted by his contemporary land-surveyors. Inō was quite proud of it himself.

4 The Drawing of the "Combined" Maps and District Maps

(*A*) *The "Combined" Maps*

When a surveyed route was divided into small sections of from 5 km. to 10 km., as described in the preceding paragraphs, and their drawing and correction were completed, such small-section maps were pieced together to make a map of an area of which the surveyed route extended to from 60 km. to 80 km. —namely, an area to be projected on a paper about 2 m. in length and about 1 m. in width, on a scale of 1:36000. Such a map was called a *yosedzu* or a "combined or joint map."

The method Inō adopted in piecing small-section maps into a "combined" map may be studied practically from the various maps still extant. On the *nishinouchi* paper, pieced together to be as large as 2 m. in length and 1 m. in width, some white lines parallel to one another from 6 cm. to 9 cm. apart were marked lengthwise (or sidewise) and these were supposed to be in the N. S. (or the E. W.) direction. The lines perpendicular to them were considered to be in the E. W. (or the N. S.) direction, similarly as in the case of a small-section map. On a "combined" map representing such an area as extending more than a half degree, Inō still treated the surface of the earth as a plane; in projecting it on a plane, he considered it possible to represent the true configuration of the meridians and parallels by two series of equidistant parallel straight lines intersecting at right angles. In collecting small-section maps on the drawing-paper, a convenient point was chosen on the paper as the starting point, from which were measured the revised E. W. component length x' and the N. S. component length y' of the first small-section map along the E. W. and N. S. lines respectively; and another terminus of the first small-section map was determined. From this point was then measured the revised E. W. and N. S. component lengths of the second small-section map, to determine the terminus; the process was repeated until the terminus of the last small-section map to be collected in the "combined" map was determined; and then the surveyed route connecting these termini of each small-section map and the positions of the various surveyed objects were transferred from the revised small-section maps to the drawing-paper by piercing it with a needle-point. And when such a measure was taken, the azimuth lines towards the distant mountains and islands were each drawn, according to the "Santō-hōiki" (The Record of the Azimuths of Mountains and Islands), with straight lines from the various points of observation, and the positions of the objects were thus determined by the intersections of the azimuth lines towards the same objects, which at the same time was intended to be of service in making corrections.

However, the effect of Inō's ignoring the curvature of the earth in drawing the "combined" maps could not be so slight as in the case of the small-section maps. The relative position of a given point to another on the map would naturally be different in accordance with the different form of the surveyed line connecting the two points; consequently, the azimuth lines drawn with straight lines from different stations on the surveyed line towards the same mountain in the distance were not to converge on one point even though no observation error was introduced. Moreover, as the magnitude of divergence caused by the slightest error in azimuth measurement magnified itself as the distance was increased, the azimuth lines drawn by Inō on the "combined" map from different stations towards a mountain or island at the distance of 40 km. or so were generally found to cross one another promiscuously within a range of from 1 cm. to 2 cm. This rendered it practically impossible to correct the position of the surveyed line by this result. Thus, except when the presence of a serious mistake in surveying was apparent in the greatness of divergence, he never made use of this in correcting the position of the surveyed line. He simply determined the position of the object sighted from a distance at the approximate centre of the portion variously intersected by the azimuth lines. On the other hand, the azimuth lines drawn from the surveyed line towards a comparatively near object generally did not present much discordance at the converging point through slight error in observation; and the observation-materials of this kind were frequently used, as in the case of small-section maps, in examining for a mistake in surveying and in correcting the relative position of adjoining small sections. By such adjustments, the drawing error due to the assuming of the area of the "combined" map as a plane had naturally a tendency to be slightly decreased.

Moreover, if, in drawing a "combined" map there chanced to be another surveyed line running near the surveyed line or intersecting it, this was made use of in adjusting the map. If the accuracy of both the surveyed lines was about alike, the positions of the related points on both surveyed lines were determined by compromising the results obtained from both. When there was a considerable difference in the accuracy of the two, the materials of the survey-line with a greater accuracy were, as a rule, consulted in determining the positions of various points, while the part of the surveyed line of less accuracy was suitably modified in shape and position for adjusting the disagreement.

It is evident that when a surveyed line actually closed surrounding an extensive area, it never did so on the map, because of the faulty method of projection, even though no error was introduced in surveying. An examination of Inō's drafts still preserved shows that he had no definite rational method of adjustment in such cases. He rendered some convenient correction only on a small part of the surveyed line, taking a temporizing measure so as apparently to smooth off the discrepancy.

After the "combined" map was adjusted as stated above, the E. W. and N. S. component lengths between both ends of the surveyed line contained on the map were measured along the E. W. and N. S. lines; and their numerical values given on the map, as in the case of a small-section map. Maps of this kind were taken as the original to be put together to compose a larger one called the district map. Moreover, this original, being further adjusted on the construction of the district map, became the final draft of the large-scale map.

The final draft of the map was called *tsuki-dehon* or a piercing copy. In order to make a fair copy from this draft, it was placed on the drawing-paper and every curve in the surveyed lines, the position of every object whose azimuth had been measured, and everything else, large and small, were pierced with a needle-point through the draft. When the piercing was over, vermilion was applied to the surveyed lines according to the marks of the needle-point, and the mountains, the temples and shrines, the trees, etc. were all represented as true to their actual state as possible around the corresponding needle-marks according to the sketched picture along the surveyed routes which had been prepared by his men; the rivers, ponds, moors, farms, and forests as well as castle-towns and villages—indeed, everything that had come within their sight—were all given and described, true to their actual state; and the coast was variously coloured, designating a sandy shore or a precipitous cliff; definite symbols were respectively adopted for the position of a province boundary, a county boundary, an astronomical station, a temple or a shrine, and a port or an inlet; and names of provinces, counties, and villages, whether a territory belonged to the Shogunate Government, a Daimio, or a temple or a shrine, and names of mountains and rivers, lakes and ponds, temples and shrines, and those of the ruling Daimios;—all these were recorded minutely. In such fair copies, however, neither meridians, parallels, nor azimuth lines towards objects observed from a distance were drawn. In making several fair copies from a piercing copy in such manner, it seems that, to save the trouble of making one sheet each time, usually, two or three or even more sheets laid one upon another were pierced with a needle-point.

(*B*) *The District Maps*

The so-called district map means one covering the whole area surveyed by Inō on each of his surveying journeys. It was constructed by uniting some "combined" maps, described in the preceding paragraphs, and correcting them by a consideration of the results of astronomical measurement of latitude. If the scale adopted was 1:36000, a sheet of the "combined" map alone extended to 2 m.; and in piecing together enough sheets to make up an area covering two or three degrees it would have rendered the size of the map enormous, and caused much inconvenience in studying the map synoptically and in giving it proper adjustments. To make a district map, therefore, "combined" maps were first reduced to 1/6 (or the scale of 1:216000) with the reducing-scale and then were joined together. In making such combination and adjustment, the following method was generally adopted.

(1) The E. W. and N. S. component lengths between the stations for astronomical observation on the "combined" maps were thoroughly measured. (Because of the more or less correction made in drawing small-section maps and "combined" maps, their values were of course somewhat different from those of the E. W. and N. S. component lengths, computed as they were, only from the materials of the field-notes.) These N. S. component lengths were added up and the total N. S. component length and con-

sequently the geodetic latitude difference between the fundamental station and a station of astronomical observation were computed; and this geodetic latitude difference was compared with the astronomical latitude difference obtained through the altitude observation of fixed stars. Inō had compiled a table which was convenient for glancing at the difference of the two results and the manner in which the difference took place for the different stations. This table was used to obtain the reasonable correction for the N. S. position of each station on the "combined" maps which had been determined by geodetic materials only.

(2) The area surveyed on each of his surveying journeys was always surrounded with the closed surveyed line and had one or two large cross-surveyed routes at least. Therefore, starting from a suitable point on the surveyed line, the E. W. and N. S. component lengths given on the "combined" maps were added up, respectively, until a complete circuit along the closed surveyed line was made by reaching the starting point; and by the values of the total amount, the amount of error accumulated in surveying the closed line was detected (the values obtained by adding the E. W. component lengths, of course, included an error arising from the assumption of the surface of the earth being a plane, besides the error in surveying); a similar process was gone through on the cross-routes to detect the amount of disagreement between the related routes.

(3) In collecting "combined" maps drawn on the scale of 1:216000, the N. S. positions were chiefly adjusted by the results of astronomical observation, so that the positions of the astronomical stations on the map coincided with astronomical latitudes. As to the relative positions among some stations close to one another, the results obtained astronomically were far less accurate than those geodetically obtained; if the position of every station had been so adjusted as to coincide with the results of astronomical

observation, the relative positions of the stations lying at a short distance would have rather increased the error on that account. In making adjustments, therefore, each station was not shifted to secure its coincidence with the results of astronomical observation, but after considering the manner of the distribution of the difference between the geodetic latitude and the astronomical latitude according to the table described under (1), the whole area was divided into a number of suitable small sections including two or three, sometimes five or six, stations for astronomical observation. Treating such stations in a section as a group, shifts were made in order to bring each section as a whole into greatest possible harmony with the results of astronomical observation, while, if any crossings had been measured, their results were also consulted for making suitable adjustments. And if such disposals caused any discordance amongst the adjoining small sections, a portion of the surveyed line in the neighbourhood of the point of discrepancy was slightly modified for the purpose of smoothing out the discrepancy.

(4) Since no effective materials were available to adjust the errors in the E. W. direction, the principal process adopted was simply such as to cancel the amount obtained in (2) as the sum total of the E. W. component lengths of a complete circuit along the closing line or cross-route;—namely, in order that no break should result in the closing line or at the junction of the cross-route on the map, and that the converging state of the azimuth lines towards distant mountains might be as favourable as possible, the surveyed lines were, in most cases, gradually pressed towards the central meridian of the map as they proceeded towards the north and thus the apparent discrepancy in the E. W. direction was smoothed over.

(5) The parallels of latitude were represented on the district map on a scale of 1:216000 by parallel straight lines with an

interval of 1·692 *shaku* for a degree; and the straight line passing a particular point in the central part of the map and intersecting the parallels of latitude at right angles, represented the prime meridian, the curves connecting the corresponding division points marked on each parallel of latitude at the intervals respectively of 1·692 cos φ *shaku* represented the meridians at the intervals of one degree. In other words, this method of drawing meridians and parallels is identical with Sanson and Flamsteed's projection, according to which the conformal relation between the original shape on a sphere and its projection on a plane is seriously affected as the distance from the prime meridian in the central part of the map increases. In order to draw a district map suited for this projection, therefore, the meridians and parallels were first to be drawn on the sheet for the district map, and the "combined" maps had to be modified so as to fit the meridians and parallels of the district map before they were collected into it. But Inō, in making a district map, collected and adjusted the "combined" maps on the paper, on which only the parallel lines corresponding to parallels of latitude were drawn, by the process described under (3) and (4); and on the completion of the map the meridians were drawn according to the above-indicated method. Consequently, his correction for the E. W. direction as compared with that for the N. S. direction, not only lacked effectual materials for its determination, but involved not a few unreasonable points in its process.

(6) The district map which had been corrected was to be the draft of the so-called middle-scale map. To make the original from this draft, piercing was done with a needle-point as described under the heading of the "combined" map, vermilion was applied on the surveyed lines, and proper symbols were given for different objects, while the topographical features were represented by the use of different colours. The middle-scale map differed from the

large-scale map in that it omitted the classes of territories and the names of the Daimios, and gave only outstanding features of each locality on the routes, though the chief azimuth lines towards the mountains and islands that had been sighted from a distance were drawn in red and the meridians and parallels in black.

The foregoing gives only a rough outline of Inō's method of drawing district maps. All the district maps Inō drew, namely,

(1) the map of the Ōshū Highway and the south-eastern coast of Yezo,
(2) that of the eastern coast of the Main Island, the Ōshū Highway and the south-eastern coast of Yezo,
(3) that of the Dewa Highway, the coast of Dewa and Echigo, and the Echigo Highway,
(4) that of the coast of the Tōkaidō and the coast of the Hokurikudō,
(5) that of the Tōkai Highway and the coast of the Chūgoku,
(6) that of the coast of Shikoku and Awaji,
(7) that of the south-eastern part of Kyūshū,
(8) that of all Kyūshū,
(9) that of the eastern coast and the Seven Islands of Idzu,
(10) that of Yezo,

were by no means drawn in the same way. The date of the survey and the amount of materials used caused a considerable difference in the accuracy of the construction of the map. In the worst case of all, "combined" maps were collected exactly as they were into a district map, being given only some unreasonable modifications as to the surveyed lines at the junctional part of the "combined" maps to smooth over the discrepancies. Again, as for the map of the Seven Islands of Idzu, Inō determined the N. S. position of each island exclusively by the astronomical latitudes, and the

E. W. position by obtaining the points on the map considered the most approximate by making use of the azimuths of the islands observed from the coast of Idzu and Sagami, the azimuths of the high peaks of Idzu, Sagami, and Suruga observed from the islands, and the azimuths of the islands sighted from one another.

5 His Construction of the Maps of Japan

There are two maps which Inō and his men drew by combining some district maps:—the map of the eastern half of Japan based on the materials of 1800–1803, and the map of all Japan drawn by making use of his entire survey-materials of 1800–1816, and Rinzō Mamiya's survey-materials of Yezo. The former was completed in 1804, when Inō was in soaring spirits, while the latter was undertaken by him in his last years, and completed by his men and pupils more than three years after his death. For this reason the method of drawing these two maps will be described separately.

(*A*) *The Map of the Eastern Half of Japan*

This map covers the region from the boundary between Owari and Ise and that between Wakasa and Echizen (about 4° west of Yedo) on the west to Nishibetsu in Yezo (about 5°·5 east of Yedo) on the east and from the coast of the Tōkai district (lat. about 34°·5*N*.) on the south to the south-eastern coast of Yezo (lat. about 43°*N*.) on the north—namely, about 9°·5 in longitude and about 8°·5 in latitude. In making this map, Inō again drew meridians and parallels by Sanson and Flamsteed's system and took the meridians passing through Kuroye-chō in Yedo as the first meridian, representing it by a straight line intersecting the parallels at right angles.

Of the three district maps (2), (3), and (4), which compose

the map in question, only (2) is extant, the others being lost. It is impossible, therefore, definitely to know Inō's method of combining and adjusting these district maps. It is evident, however, that the district map (3), which was to occupy the central part of the composed map, and whose meridians were precisely similar in form to those of the composed map, was first drawn as the framework of the map; and adjoining this on the east, the district map (2) was drawn; and as for the discrepancy in the E. W. direction arising on the adjoining portion, the surveyed line of (2) was so corrected as to harmonize with the surveyed line of (3); then the district map (4) was drawn on the west of (3), and under the same principle, the surveyed line of (4) was chiefly corrected. Since the fundamental materials have been lost, there remains no reasonable basis on which to presume whether the process Inō actually took in combining and adjusting was really proper or not, except by examining whether the longitudes of various points were correctly assigned on the map. Examining now the positions of various stations, the following facts are obtained. Though the positions of the stations in the Tōhoku and Tōkai districts were admirably adjusted, the positions of the route from the boundary line between Owari and Ise to Tsuruga, forming the western border of the map, and also the positions of various stations along the coast of the Hokuriku district, included errors greater than those of surveying, proving thoughtlessness in the matter of adjustments. And as for the south-eastern coast of Yezo, the district map (2) for this part had not been adjusted well, to begin with, and, to make matters worse, in combining this part to the main district no modification whatever had been adopted, in face of the change in the form of meridians; therefore, the meridians for this part on the composed map had almost lost any significance with regard to the positions of the various stations. No doubt, the cause of such confusion was that, in projecting on a plane a tract of land on a sphere, Inō drew meridians and parallels according to Sanson and Flamsteed's system on one hand, while, on the other hand, he attempted to retain on the map the conformal relations which were inconsistent with the former. Though the explanatory notes accompanying the map indicate that the azimuth lines of the objects sighted from a distance were drawn relative to the meridians intersecting the parallels slantwise, a comparison of a practical measurement of the azimuths on the map and those described in the "Santō-hōiki" will betray the fact that the azimuth lines were mostly drawn with straight lines relative to the parallels.

(*B*) *The Map of All Japan*

This map, which was completed in 1821, a few years after Inō's death, comprised the Five Isles of Hizen (over 7° west of Kyōto) on the west, and part of Kunajiri Island in Yezo (over 10° east of Kyōto) on the east; Yakushima in Ōsumi (lat. slightly over 30°*N.*) and Hachijōjima in Idzu (lat. about 33°*N.*) on the south, and the Sōya Cape in Yezo (lat. about 45°·5*N.*) on the north:—comprising practically the whole of the Japanese territories at that time, extending as they did more than 17° east and west and more than 15° north and south. As to the method of drawing meridians and parallels Sanson and Flamsteed's system was still adopted, assuming the meridians which passes through the site of the Calendaric Office at Nishi-Sanjō-dai, Kyōto, as the first meridian, and which was intersected by the parallels at right angles; on this account, the meridians towards the eastern and western borders of the map tended to incline towards the parallels, in extreme cases, even about 7° from perpendicularity.

In making this complete map, the district map (5) including the central region of the whole country—namely, the Tōkai

Highway, the coast of Kii, and that of the Chūgokū region surveyed in 1805–1806 and revised and corrected by the materials of later surveys of the highways of the Chūgoku region, was drawn first as the central part of the complete map; then the district maps (*A*), (9) and (10) on the north-east, and (6) and (8) on the south-west, were drawn and adjusted.

Now since these district maps had been drawn so that they held approximate conformal relations in their central parts, they had to be properly distorted according to their positions on the complete map before they could be transferred upon it. A comparison of the complete map and all the district maps serving as the materials in its synthesis proves that not only these dispositions were not effected reasonably, but serious mistakes and confusion for which no reason can be found were introduced in the adjustment.

Since the meridians of the district map (5) like those of the complete map had been so drawn that one of them passing through the site of the Calendaric Office in Kyōto intersected the parallels at right angles, it is quite reasonable that, in transferring (5) to the complete map, Inō adopted it in its entirety, without modifying its shape except by adding some newly surveyed highways to revise it. Despite the fact that the meridians of (*A*), the map of the eastern half of Japan, were so drawn that one of them passing through Kuroye-chō in Yedo intersected the parallels at right angles and differed in form from the meridians of the complete map, the map (*A*) was drawn whole on it, relying for the positions of the various points on the Tōkai Highway of the map (5), only revising it according to the new materials obtained by surveying the Nakasen Highway and other highways, and disposing of the discrepancy arising on the boundary line between Wakasa and Echizen Provinces, simply by distributing it along the coast-line between the boundary and Echigo. And as for the points along the eastern coast of the Main Island and in the Ōu district, those of the map (*A*) were transferred whole with no modification whatever upon the complete map. As his method of composing was so extremely unreasonable rather than thoughtless, the various points correctly drawn regarding their E. W. positions on the map (*A*) came to show a considerable drawing error if referred to the meridians of the complete map. The greater became the amount of the drawing error, on the coast of the Hokuriku district, the further north-east it progressed, until it was already from 4′ to 7′ on the coast of Noto and Echigo. On the eastern coast and in the Ōu district, the drawing error became all the greater the further north it progressed, until in the vicinity of Mimmaya or Noheji the error increased so greatly as to be about 20′.

And in transferring the district map (9) including the eastern coast of Idzu and the Seven Isles, no correction whatever necessary for the difference of the shape of the meridians was given; it was added to the complete map on the immediate south of Sagami and Suruga which had already been drawn on the complete map, with the result that, referring to the meridians of the complete map, Hachijōjima had an error of about 10′ in longitude. The district map of Yezo (10) was also transferred entire with no modification, except determining the E. W. position of the coast-line in the neighbourhood of Matsumae and Hakodate, so that it harmonized with the northern coast-line of Mutsu of the map (*A*) which had already been transferred. Since meridians on the complete map suffered the greatest distortion in the district of Yezo, the drawing error of the longitude, referred to them amounted to more than 20′ in the neighbourhood of Hakodate, and increased gradually as it progressed further north in the island of Yezo till it grew over 30′ in the neighbourhood of Otaru, and so serious as to amount to about 50′ in the neighbourhood of Sōya, the northern end of the island.

And in drawing the district map of Kyūshū (8) on the immediate west of the central part, the original was also transferred entire on the complete map, merely determining the positions of the connected parts so that the coast-line of Buzen and Bungo in the district map of Kyūshū approximately harmonized with the coast-line in the neighbourhood of Nagato, the western end of the central district. Thus, the meridians on the complete map naturally lost their proper significance for the Kyūshū district also, and though in the neighbourhood of the Shimonoseki Strait they served to give an approximate value, the drawing error gradually increased as it progressed further south until it amounted to somewhat less than 4′ in the neighbourhood of Nagasaki, and 7′ or 8′ in Satsuma Province.

In adding the district map of Shikoku (6) to the complete map, it was simply transferred entire on the south of the Chūgoku district and between the Kii Peninsula and Kyūshū; but since this district had survey-materials mutually related with a number of islands belonging to the Sanyō district all along its northern side, and also those mutually related with various points in Kyūshū on the west, the discrepancy arising between them was adjusted, relying chiefly upon the positions of the district maps of Chūgoku and Kyūshū, and modifying the district map of Shikoku. On this account, the western part of Shikoku was displaced in the direction opposite to that properly required for modifying the meridians. However, as this district is not far from the central part and consequently the change of the form of the meridians is not considerable, the longitudes of various points referred to the meridians of the complete map are not affected very much by the drawing error.

Thus, the complete map of Japan embodying Inō's monumental work includes very serious drawing errors and does not represent suitably the results of his surveying. On the whole, it is hardly more than that the various district maps projected according to the meridians differing in form from one another were plotted on one sheet thoughtlessly, and meaningless meridians were drawn upon it.

6 The Extent of His Knowledge of Map-Projection

As stated in the preceding paragraphs, although Inō's method of adjustment, adopted in projecting small-section maps covering a few km., was quite reasonable, his method of harmonizing and adjusting came to lack exactness in drawing the "combined" and the district maps, and his projection of the eastern half of Japan betrayed far greater lack of caution. This reveals that Inō had no profound knowledge of projecting a tract of land on a sphere upon a plane. As for his serious errors in the complete map of Japan, however, it is difficult to explain them away simply on this basis, and this requires a special consideration.

Whether or not the method of drawing meridians and parallels adopted by Inō in his mapping was followed to a method introduced from abroad is not clear. When we recollect that Inō's measuring-instruments, his method of measurement, and the principles of his map-drawing were established under the direction of Yoshitoki Takahashi, it is evident, however, that this method of drawing meridians and parallels was proposed by Takahashi. The reason why he adopted a method equal to the so-called Sanson and Flamsteed's projection was because it is so simple in drawing and because, as the parallels of latitude are represented by parallel straight lines, it was somewhat convenient in correcting for the N. S. positions of various points on the map by astronomical materials.

The disadvantages and inconveniences of this projection through lack of conformal relations are more numerous than can be cited, and the maps drawn on such a projection are necessarily unsuited for general purposes. It was, indeed, soon after Inō commenced his surveying that Yoshitoki Takahashi became aware of the defects, and eagerly set about improving the method of drawing meridians and parallels or map-projection until he devised a plan which he still assiduously improved with a view to ask Inō to make maps by the new plan representing the topography properly as soon as he had finished his surveying. On April 8th, 1803, Takahashi wrote a letter to Shigetomi Hazama, telling him that, since it was improper to represent the parallels of latitude by straight lines as had been done previously, he desired to introduce curves instead, but that Inō did not like them because the projection involved much trouble and that he would make Inō adopt the new method in drawing the map of the eastern half of Japan. Takahashi also informed Hazama of the new method he had designed and asked for the latter's opinion about it. Takahashi's plan introduced in the letter was as follows:—

In drawing meridians and parallels which suit a tract of land extending from lat. $\varphi_1°$ to lat. $\varphi_2°$ and from long $\lambda_1°$ to long. $\lambda_2°$, a trapezoid $CDEF$ with $AB=k(\varphi_2-\varphi_1)$ for the height,

$$CD=2\,k \tan^{-1}\left(\tan\frac{\lambda_2-\lambda_1}{2}\cos\varphi_1\right)$$

for the base, and

$$EF=2\,k \tan^{-1}\left(\tan\frac{\lambda_2-\lambda_1}{2}\cos\varphi_2\right)$$

for the upper side is first drawn, k representing the length of one degree of arc on the surface of the earth in a suitably reduced scale. Then with the point O, on which the two sides CE and DF meet, as its centre and OA as its radius, the arc $C'AD'$ is drawn to represent the parallel of latitude φ_1; and with OB as its radius, the arc $E'BF'$ to represent the parallel of latitude φ_2; the

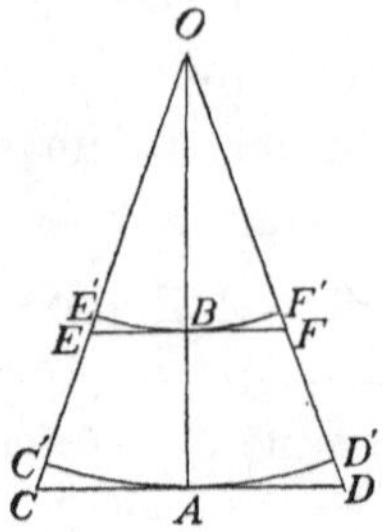

Fig. 46.

meridian of λ_1, is represented by the straight line $C'E'$, and the meridian of λ_2 by the straight line $D'F'$; the parallels of latitude between φ_1 and φ_2 are represented by the concentric circles drawn through the points dividing the straight line AB into equal parts, and the meridians between λ_1 and λ_2 are represented by the straight lines connecting the points dividing the arcs $C'D'$ and $E'F'$ into equal parts.

This new projection which Takahashi devised is somewhat similar in form to a combination of Soldner's projection and the simple conic projection. Its magnification is approximately constant for a tract of land covering 10° or so both in length and width, and does not lose the conformal relations to any considerable extent; the azimuth lines towards distant objects, even if all represented by straight lines, usually do not cause an error greater than 15′. Therefore, this might be called a suitable projection for practical use, when compared with the accuracy of the surveying of these days.

In view of the fact that the new projection Takahashi devised was of far greater advantage in practical use than Sanson and Flamsteed's system, it is quite strange that Inō should still have adopted the same old system with the map of the eastern half of Japan completed in September, 1804. The reason must be that the death of Takahashi in February of the same year and Inō's reluctance to take greater trouble for projecting by the new method impelled him to act expeditiously. This at the same time proves Inō's lack of knowledge concerning map-projection, and his inferiority in scholarship to Takahashi. It follows, therefore, that Inō himself is responsible for marring the accuracy of his maps as a consequence of his imprudence in projecting and adjusting the map of the eastern half of Japan.

Inō consistently adopted Sanson and Flamsteed's projection for drawing all his district maps until his last years. It is doubtful, however, whether he depended upon the system with a full realization of the nature of the projection, and willingly sacrificed part of the conformal relation. As already stated, he, in drawing the azimuth lines on the map of the eastern half of Japan, paid no particular attention to the slantwise intersections of the meridians and parallels. This leads us to suspect that he thoughtlessly drew meridians and parallels by Sansom and Flamsteed's system, without carefully considering the effect of the slant intersections of meridians and parallels. And the fact that the same method was adopted even on the complete map of Japan, finished after Inō's death, when a serious error of drawing was committed, may tend to prove that he died without fully understanding map-projection. There remains, however, counter-evidence strong enough to defend him on this point.

The counter-evidence is the existence of a manuscript book entitled "Ichiri-rokubudzu-tōzai-no-keido-narabini-hokkyoku-kakoku-yori-no-chokuyenkeisa" (The E. W. Co-ordinates and N. S. Co-ordinates on the Map Drawn on the Scale of 1:216000), in which Inō computed the rectangular co-ordinates for the points of the intersection of meridians and parallels according to a new system of projection which he, dissatisfied with Sanson and Flamsteed's method, devised as a far more suitable one than Takahashi's new plan. Considering the fact that Kagesuke Shibukawa mentions this work in his biography of Inō as one of the books Inō wrote, and also considering its contents, it is evident that Inō himself computed the values after finishing his field-work with a view to practical use.

In this book, the latitudes are given in the interval of one degree ranging from 29° to 44°, and the longitudes in the interval of 10′ ranging from 0° to ±8° to be available for any point in

Japan. And the numerical values of x and y co-ordinates of the point, the latitude and the longitude of which are respectively φ and λ, are tabulated, the central meridian being taken as the axis of y, and the line drawn to the central meridian perpendicularly through the point of the intersection of the central meridian and the parallel of latitude φ as the axis of x. These values of x and y are given in the reduced scale of 1:216000 and are computed by the following equations:

$$x=\lambda \cos \varphi \times 28{\cdot}2 \times 6$$

$$y=\left[\tan^{-1}\left\{\frac{\tan (90^\circ-\varphi)}{\cos \lambda}\right\}-(90^\circ-\varphi)\right]\times 28{\cdot}2 \times 6$$

where λ and φ are represented in degrees, and the unit of x and y is *bu* or 1/100 *shaku*. It may be seen from these equations that x is the length (represented in the reduced scale) of the parallel of latitude φ for the longitude difference λ; and y is the length (represented in the reduced scale) on the earth's surface corresponding to the difference between the latitude φ and the latitude of a point at which the meridian of the longitude λ intersects the great circle drawn perpendicularly to the central meridian through the point of the latitude φ on the central meridian. In other words, referring to the diagram, if PZ be the central meridian, PC the meridian of the longitude λ, P the north pole, Z the point of the latitude φ on the central meridian, ZA the small circle representing the parallel of latitude φ passing through the point Z, and ZC the great circle perpendicular to PZ at the point Z, then x for the point A of the latitude φ and the longitude λ is found by the length of the arc ZA, and y by the length of the arc CA, both being represented in the reduced scale of 1:216000. In short, this new projection may be obtained by shifting the points of the

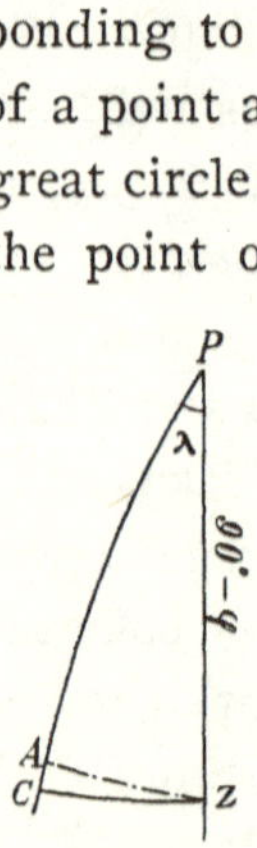

Fig. 47.

intersection of the meridians and parallels drawn by Sanson and Flamsteed's projection toward the north by the length of CA corresponding to every point of intersection. Now taking the central meridian represented by a straight line as the axis of Y and the straight line perpendicularly intersecting it at lat. 35° as the axis of X, and assuming the earth to be spherical, the numerical values of the co-ordinates for various stations lying near the frontier of Japan were computed anew by the simple conic projection. Bonne's projection, Soldner's projection, and the conformal projection obtained by modifying Soldner's in order to compare with those by Inō's projection and Takahashi's projection. Their results are shown by the table on page 286. (For Takahashi's projection, the lines of $\varphi=29°$, $\varphi=41°$, $\lambda=\pm8°$ are taken as the boundary lines. For the simple conic projection and Bonne's projection, the parallel of latitude along which the cone touches is taken to be 35°. The length of one degree of the meridian arc is taken as 28·2 *ri* and represented by the reduced scale of 1:216000.)

As is shown by this table, Inō's new projection gives the values of the co-ordinates more approximate to the conformal projection than any other within the range in question. The difference between Inō's projection and the conformal may be given by the following approximate equations.

$$Y_{(\text{conformal})} - Y_{(\text{In\=o})} \fallingdotseq 28{\cdot}2 \times 6 \times \left(\frac{\pi}{180}\right)^3 \frac{(\lambda^\circ)^4}{16} \sin 4\varphi$$

$$X_{(\text{conformal})} - X_{(\text{In\=o})} \fallingdotseq 28{\cdot}2 \times 6 \times \left(\frac{\pi}{180}\right)^2 \frac{(\lambda^\circ)^3}{6} \cos\varphi \cos 2\varphi$$

These show that the difference of X reaches 0·02 *shaku* in the extreme case of $\varphi=29°$, $\lambda=8°$, while the maximum difference of Y is only 0·002 *shaku*. Thus, in spite of the fact that the new projection Inō designed had no particularly important mathematical property, it coincided closely with the conformal projection

			Conformal Projection	Inō's Projection	Takahashi's Projection	Simple Conic Projection	Bonne's Projection	Soldner's Projection
			shaku	*shaku*	*shaku*	*shaku*	*shaku*	*shaku*
Y	$\lambda=8°$	$\varphi=29°$	− 9·749	− 9·751	− 9·683	− 9·676	− 9·681	− 9·749
		$\varphi=35$	+ 0·446	+ 0·445	+ 0·437	+ 0·444	+ 0·444	+ 0·446
		$\varphi=41$	+10·622	+10·621	+10·558	+10·563	+10·558	+10·622
		$\varphi=44$	+15·702	+15·702	+15·618	+15·623	+15·613	+15·702
	$\lambda=6°$	$\varphi=29°$	− 9·926	− 9·927	− 9·888	− 9·884	− 9·887	− 9·926
		$\varphi=35$	+ 0·250	+ 0·250	+ 0·246	+ 0·250	+ 0·250	+ 0·250
		$\varphi=41$	+10·416	+10·416	+10·330	+10·383	+10·381	+10·416

			Conformal Projection	Inō's Projection	Takahashi's Projection	Simple Conic Projection	Bonne's Projection	Soldner's Projection
			shaku	*shaku*	*shaku*	*shaku*	*shaku*	*shaku*
X	$\lambda=8°$	$\varphi=29°$	11·859	11·839	11·820	11·888	11·826	11·830
		$\varphi=35$	11·100	11·088	11·016	11·076	11·076	11·076
		$\varphi=41$	10·220	10·216	10·212	10·264	10·205	10·201
		$\varphi=44$	9·738	9·737	9·810	9·858	9·727	9·722
	$\lambda=6°$	$\varphi=29°$	8·888	8·879	8·869	8·921	8·874	8·875
		$\varphi=35$	8·321	8·316	8·266	8·311	8·311	8·311
		$\varphi=41$	7·664	7·662	7·663	7·702	7·657	7·656

and it may be said to satisfy practically the relation of similarity which is very desirable in projecting a map. This projection, however, was attained by chance as a result of comparative study of Takahashi's projection with Sanson and Flamsteed's projection which had always been adopted. It should not be supposed that this was obtained by a fundamental study concerning the conformal relation, because this is against all evidence, in view of the stage of the development of the science in Japan at that time, and also because of the circumstances of the period.

Although Inō came upon a new projection in his last years and eagerly attempted to use it in practice, he noticed that a district map extending over two or three degrees would make little difference in its shape, no matter whether the old projection or the new was adopted. Therefore, it was his desire to project district maps temporarily by the old method, and to revise them by the materials of later surveys, modify and adjust them slightly so that they might suit the meridians and parallels of his new projection, and then transfer and join together these district maps on a sheet of paper on which the meridians and parallels of the new system had been drawn. With this purpose, he set about revising district maps. But before his task of revising the district maps was accomplished, and consequently before he had set about compiling all the maps, he succumbed to his last illness. His ambition was not fulfilled after all, and the result was serious, as hitherto indicated. It resembles the previous case when Inō drew the map of the eastern half of Japan without following Yoshitoki Takahashi's advice.

Thus it may not be said that the great mistake shown in drawing the complete map of Japan, finished in 1821, was caused by Inō's imprudence. The petty officials who were personally in charge of the work were only men who obeyed the rule, and did merely what they were told. Such men, with no profound knowledge, could not be held responsible for the blunders. The one man who could not possibly evade responsibility for this was Kageyasu Takahashi, son of Yoshitoki Takahashi, who superintended the work of surveying and map-construction. Young Takahashi not only supervised Inō's surveys, but was himself an accomplished scholar of no mean ability in the field of geography. An examination of the maps he drew about 1810 shows that he adopted the stereographic projection for the map of the whole world and one similar to the conical projection for the sketch-map of the Japanese frontiers—the projections suitable to the respective cases. The fact serves to prove that his knowledge of map-projection was by no means meagre. If young Takahashi had not left the work of map-construction with his petty officials after Inō's death and had himself carefully supervised it, the serious mistakes shown on the complete map of Japan would not have occurred. After 1813, when a fire destroyed the Astronomical Bureau and young Takahashi lost in it the manuscript of a book on the Manchurian language, on which he had spent years of labour, he was so whole-heartedly devoted to the re-writing of this book that he paid almost no attention to anything else, and neglected his duties in supervising map-construction. So it was that he overlooked their outstanding blunders, or noticed them only too late to effect a remedy.

We may conclude that, for the improper adjustment and compilation of the map of the eastern half of Japan, Inō himself was responsible; but for the great mistakes in projecting the complete map of Japan, Kageyasu Takahashi was principally responsible—responsible at least for the remiss control of his subordinates. Or, it may be said that for the former the death of Yoshitoki Takahashi was the serious indirect cause, while for the latter that of Inō himself was.

CHAPTER VI

THE ACCURACY OF HIS SURVEY AND OF HIS MAPS

In considering the accuracy of the results of Inō's survey, abundant materials exist for a concrete investigation of the accuracy of some of their elements, such as latitude and azimuth measurements, while there are no authentic fundamental documents for other elements, such as distance measurement, whose accuracy can only be inferred from some fragments. We shall first discuss the accuracy of these elements respectively and then deal with that of the measurement made up from them and of his maps.

1 His Distance Measurement

Instrumental Errors. In order to investigate the accuracy of the distance measurement Inō carried out with various measuring-ropes in expectation of high exactitude, we have first to determine the accuracy of the length of the measuring-ropes he used. As for the standard *shaku* scale upon which the length of the measuring-rope depends, it must be admitted that Inō disregarded an error of about 1/3000 of its length, as already noted in the paragraph on the scale. Consequently, the measuring-ropes standardized by it—even the iron chain which was supposed to be the most accurate of all the measuring-lines, should be considered to include an error of this amount at least. Though the bamboo-ropes and rattan-ropes were not so greatly affected by humidity as the hempen ropes, they sometimes suffered considerable variation, and those made slender for convenience in use changed their length even with a slight difference in tension;

therefore, their accuracy, perhaps, failed to exceed a half of that of the iron chains. It is supposed that the hempen ropes, even when standardized before and after the measurement, had very little accuracy in practice.

Observation Errors. Again, the observation error introduced into a distance measurement with these measuring-ropes, of course, varied according to the kind of ropes, the state of the measured route, and the condition of the weather. Though materials are not extant for definitely determining the degree of their accuracy, it is presumable that if 10 km. or so of a fairly even road in good weather was measured with an iron chain in 7 or 8 hours by the method Inō adopted, an error as great as 1/2000 could not be avoided; if the routes were much curved, rough, or ankle-deep in mud after a rainfall, the error was no doubt increased considerably. Where the bamboo-rope, the rattan-rope, or even the hempen rope was used, it is impossible to distinguish clearly an error of observation from that of the instrument, and taking these two kinds of errors together into consideration, it is fairly correct to suppose that the use of a bamboo-rope or a rattan-rope caused an error about twice as great as in the case of using an iron chain, and that the use of a hempen rope introduced a still greater error.

The accuracy of using a perambulator can be investigated from some extant documents, according to which the instrument generally introduced an error of 2/1000 or 3/1000 and sometimes 6/1000 or 7/1000 in the measurement of a distance of about 5 km.

The distance used in drawing a map was the horizontal length obtained by multiplying the cosine of the inclination of the measured line into the measured distance. So the accuracy of the horizontal distance may be somewhat decreased by reason of an observation error in the inclination. However, the extreme smallness of the inclination of the measured line kept its influence generally to a minimum.

Some of the errors are systematic by nature, and some are accidental; the former if expressed in percentage is nearly constant for the length of the measured line, while the latter, if expressed in percentage, decreases as the length of the measured line increases; thus, in the distance measurement the percentage of its total error will decrease as the distance increases. Of the two kinds, however, what may be considered as the systematic error was comparatively prevalent. Besides, the increase of distance involved the frequency of unfavourable weather and the ruggedness of the measured route which considerably marred the accuracy of the measurement. It seems, therefore, that the percentage of error in Inō's distance measurement did not always decrease with the increase of distance.

For a long route, Inō seldom succeeded in keeping the error of his distance measurement below 1/1000 of the distance measured. Such accuracy was secured perhaps only on the Ōshū Highway from Yedo to Shirakawa and on the Dewa Highway from Shirakawa to Mimmaya—the route measured with the utmost care, for the purpose of determining the length of one degree of the meridian arc. Besides, since part of the coast-line of Dewa and Echigo, the Tōkai Highway between Yedo and Kyōto and Ōsaka, and the Sanyō Highway between Ōsaka and Shimonoseki, were not only quite smooth, but were very likely measured by Inō with special care for determining the length of one meridian degree or computing longitude, these measured routes may come next in accuracy; the error may not have exceeded 1/500 at the most. Despite his great care in measurement, however, it is evident that the ordinary coast-line could not approach these standard routes in accuracy on account of the greater ruggedness of the land.

2 His Azimuth Measurement

Instrumental Errors. In order to consider the accuracy of the measuring-instruments which Inō used in azimuth measurement, we shall first examine the graduated circles. They varied in accuracy: the smallest theodolite was graduated to each degree on a circle 10 cm. in diameter, while the largest was so minutely graduated as to each 5′ on a circle about 75 cm. in diameter. And their graduation was so finely executed that the error is believed to have been not more than 0°·1 on one with the smallest diameter, and less than 1′ on the largest instrument.

Let us now consider the relation between the alidades and the graduated circle. As for the small magnetic theodolite graduated in the reversed order, whenever the magnetic needle and the pivot were repaired, the alidades were removed from the instrument, but in re-installing them, they were adjusted by stringing a fine line temporarily through their fine holes and by keeping the line parallel with the meridian of the graduated circle. Since this adjustment was made only with the naked eye, though Inō claims that the greatest care was taken, it was inevitable that an error of 0°·1 or so should have been introduced. On the large and the middle-sized magnetic theodolite and the semi-circular magnetic theodolite, the above relation was changed because of the difference in construction, to the parallel relation between the line of sight and the line of the index arm; and the error introduced, as these two failed to be perfectly parallel with each other, certainly exceeded the graduation error of the circle.

As regards the semi-circular magnetic theodolite, the effect of eccentricity should be considered. When we study the extant instrument, the axis of the index plate is very elaborately constructed to enable the index plate to rotate smoothly about the centre of the graduated circle; still it appears impossible to make the axis of rotation and the centre of the circle always coincide with each other, even within the range of 0·3 mm., because of the presence of some eccentricity and play. If an eccentricity of 0·3 mm. is supposed to exist at the centre, the maximum error caused by this on the large semi-circular magnetic theodolite about 15 cm. in radius will reach 7′, and 8′ on the small semi-circular magnetic theodolite about 12 cm. in radius—the values several times as great as the smallest unit of angle measurable with the graduated circles.

The next question is as to what degree of accuracy Inō's magnetic needle indicated the direction of the magnetic meridian. The magnetic needle used by Inō was always in the form of a needle, and it is clear that there existed no great difference between the magnetic axis and the axis of its geometrical form; but it cannot be said that the two always coincided exactly with each other. Moreover, the needle form of the magnet decreased the magnetic moment and consequently caused the disadvantage of affecting the pointing direction of the magnetic needle even with the slightest friction at the pivot, and this of course introduced some error. With a view to gaining an idea of the amount of the error, a magnetic needle Inō had used was carefully straightened and strongly magnetized afresh, and then suspended upon a sharp-pointed pivot for experiment; it appears impossible to avoid an error of 0°·1 or so. As the construction of the magnetic needles did not vary according to the size of the theodolites and their length was usually 10 cm. or so, the amount of the error in question should be regarded as almost equal, regardless of the size of the theodolites.

To put together the various instrumental errors, it seems that, in spite of the minute and accurate graduation up to 1′, the large and middle-sized magnetic theodolites failed to exercise their full capacity checked by the imperfect accuracy of the magnetic needle as their orientation was adjusted by the magnetic needle; and they could not be expected to attain an accuracy exceeding 0°·1 even under the most favourable conditions. The semi-circular magnetic theodolite, more subject to the error arising from eccentricity than the large and middle-sized theodolites, of course attained less accuracy. Compared with the amount of other errors, however, it may be said that this eccentric error inflicted no serious effect upon the accuracy of the instrument. The small magnetic theodolite, of course, could not approach the large theodolites in the accuracy of graduation on the circle. But since the errors arising from other parts were not much greater than those of the larger theodolites, an accuracy of 0°·2 or so could be expected from the small magnetic theodolite, where an accuracy of 0°·1 or so was expected from the larger instruments.

Observation Errors. When an observation was made with a small magnetic theodolite graduated in the reverse order, the circle was moved with the alidades until the azimuth of the object was read off by means of the magnetic needle. When using the large, the middle-sized, or the semi-circular magnetic theodolite, the instrument was first set up by the magnetic needle and the observed azimuth was read off by the index arm attached to the alidades; thus, the method of observation varied. Of all the various observation errors, that introduced in reading the azimuth by means of the point of the magnetic needle on the small magnetic theodolite is similar in nature and amount to the error introduced in keeping the direction of the magnetic needle and the meridional line marked on the circle identical for setting up the instrument, in case one of any other type was used. Experiments on the instruments Inō had used proved the amount of this error in either case to be 0°·1 or 0°·2. Then, as for the observation error introduced in sighting the object through the alidades, the error on the large and middle-sized magnetic theodolites provided with telescopes was perhaps small enough to be ignored in comparison with other errors. But it was different with the alidades for naked-eye observation. The amount varied according to the length between the slits of the alidades and the skill of the observer, but it may be justly estimated at 0°·1 or 0°·2. Finally, the error introduced in reading the graduation by the index arm on the large, the middle sized, or the semi-circular magnetic theodolite was so small as to be ignored, compared with other errors.

Thus the only disadvantage the small theodolite with the circle about 10 cm. in diameter had against the large theodolite 75 cm. in diameter and provided with a telescope, was the comparatively large sighting error. In other errors, there was no remarkable difference between this small instrument and the larger ones. However, since the large, the middle-sized, and the semi-circular magnetic theodolites were securely fixed upon the stout stands when used, they were fit for careful observations, while the small theodolite was not fixed, but suspended on a universal joint, and reading of the graduation by the point of the magnetic needle more or less in motion, naturally gave rise to some disadvantages. The total observation error, therefore, may be estimated at 0°·1–0°·2 on the large magnetic theodolite and 0°·2–0°·3 on the smallest magnetic compass mounted on a stick.

To put together both the instrumental and observation errors investgated heretofore, we may conclude that it was difficult to expect an accuracy of over 0°·2 in the observed azimuth even with the largest magnetic theodolite, while it was not very difficult to expect an accuracy of 0°·3, differing little from the other, even with the smallest magnetic theodolite. The reason why Inō neglected the large theodolite and used the small one in spite of Takahashi's preference was his recognition of this fact as a result of the practical experiments with various magnetic theodolites in his field-work.

On the other hand, with the assumption that the accuracies of the middle-sized, the small, and the semi-circular magnetic theodolite were approximately equal, and also the instrumental and observation errors introduced in azimuth observations with these instruments arose according to the law of errors, the values of the observed azimuths of various objects sighted with various magnetic theodolites at each station mentioned in the "Santō-hōiki" may be utilized here to compute the mean error for a single observation by the method of least squares. For instance, some thirty stations were picked up at random from the extant 67 volumes of the "Santō-hōiki" and, making use of the materials of the azimuth observations at each station, the mean error for a single observation at each station has been tentatively computed. The following values which do not differ much from one another were obtained.

±19′·4, ±11′·4, ±14′·6, ±11′·2, ±8′·0, ±8′·5, ±11′·2, ±9′·3, ±10′·7, ±7′·2, ±18′·8, ±16′·0, ±14′·3, ±12′·9, ±10′·5, ±16′·4, ±11′·3, ±13′·2, ±14′·1, ±13′·7, ±15′·3, ±12′·2, ±14′·4, ±17′·1, ±16′·2, ±14′·4, ±11′·3, ±12′·8; the average of these values being ±13′·1.

The amount of the mean error computed from the results of observations by the law of errors thus very closely coincides with that directly inferred from the construction of each instrument. The value here obtained, or 13′ may, therefore, be taken as correctly indicating the mean error for a single azimuth observation.

Once the mean error of a single observation is estimated to be about 13′, the mean error of the azimuth of each segment of the surveyed line, as it was generally determined by comparing the results of fore-sighting and back-sighting measured with two magnetic theodolites, except in the earliest stage of his career, should have been

$$\frac{\pm 13'{\cdot}1}{\sqrt{2}} = \pm 9'{\cdot}3,$$

namely, about 10′; and since, in measuring the azimuths of mountains and islands, more than two, sometimes six or seven—generally about three—magnetic theodolites were used, the mean error of the azimuth measurement of these mountains and islands should have been from

$$\frac{\pm 13'{\cdot}1}{\sqrt{2}} = \pm 9'{\cdot}3 \quad \text{to} \quad \frac{\pm 13'{\cdot}1}{\sqrt{7}} = \pm 5'{\cdot}0,$$

namely, about 7′ or 8′ in most cases.

Magnetic Declination. What has been discussed in the foregoing paragraphs all relates to the errors of the azimuth referred to the magnetic meridian, but not those of the true azimuth. As a result of his own experiments, Inō denied the existence of the magnetic declination, attributing such deviation to the poor quality of the magnetic needle. As a matter of fact, however, the magnetic meridian did not exactly coincide with the true meridian, and, moreover, the declination varied at different hours of the day because of the so-called diurnal variation. It is thus necessary to consider the influence of the magnetic declination and its diural variation in order to obtain the error for the true azimuth.

Let us first consider the diurnal variation. No material is extant for studying the actual state of the diural variation in Japan at Inō's time. According to the result of the magnetic survey of Japan made about 1895, some 90 years after the period, the amount of diurnal variation was greatest in summer, the whole amplitude being from about 5′ to 10′, and the variation was most active from 8 a.m. to 3 or 4 p.m. If the state of the variation had been nearly similar at Inō's time, he had measured azimuth during the hours of the day when the variation was most active, because Inō was generally engaged in land-surveying from sunrise till about 3 in the afternoon; therefore, there would have been a difference of 5′–10′ between the azimuths measured early in the morning and those measured in the afternoon, and the azimuths measured in the forenoon would generally have been too small (if the azimuths had been measured in the order of N.E.S.W.) while those measured in the afternoon would have been too great. Thus, even if the daily mean value of the magnetic declination had been zero at Inō's time, the measured azimuths would have been influenced by the diurnal variation, and the error for the true azimuth would have been naturally more or less greater than that for the magnetic azimuth. The value 10′ previously obtained as the mean error of the azimuth of each segment of the surveyed line should be increased to 12′.

Again, as to the manner of the distribution of the magnetic declination itself in Japan at Inō's time, if we compare the magnetic azimuth measured for various objects at each station recorded in the "Santō-hōiki" which is still preserved, with the respective true azimuth accurately measured to-day, this may be determined definitely within the accuracy of Inō's measurement. This investigation would be extremely interesting. As it is exceedingly difficult to locate the stations at which Inō made observations more than a century ago, we shall only roughly discuss here the amount of the magnetic declination in Japan at Inō's time.

According to the "Santō-hōiki," there was always a difference of 10′–20′ among the azimuths indicated by various magnetic theodolites Inō used in his surveys of 1802. Despite this fact, in a letter he wrote to Yoshitoki Takahashi from a field-station, he says, "The six magnetic theodolites I used did not differ in their indication, to my great relief." From these words it is apparent that a difference of this amount did not trouble his mind. Hence his statement that the magnetic meridian and the true meridian coincided should be understood to mean a coincidence somewhat like this.

To find out the general state of the magnetic declination in eastern Japan at Inō's time, it would be simplest to measure afresh the true azimuth at the site of Inō's former residence at Kuroye-chō, Fukagawa, which abounds in the materials of azimuth measurement, and whose position is defined, and then compare the new results with the old. When the magnetic azimuths of Mt. Fuji, and the two peaks of Mt. Tsukuba given in the extant records were compared with the true azimuths for these points obtained from new materials, the north end of the magnetic needle in each case proved to deviate slightly east of the true azimuth, and the mean value obtained was magnetic declination $\delta=19'$E. for the year 1802. Though this value includes the observation error and the infiuence of the diurnal variation and, therefore, cannot be relied upon too minutely, it may be surmised from this that in the earlier stages of Inō's survey the magnetic north in Yedo deviated slightly east from the true north.

Again, in order to learn the general state of the magnetic declination extending over a wide area, the district map of Shikoku drawn by Inō from the survey-materials of 1808 and that of south-eastern Kyūshū drawn from the materials of 1810 were examined. On the surveyed route running north and south, the position of the southern stations, when compared with those of the northern stations, were discovered to deviate, in most cases, slightly west from the true positions. And when the materials for the astronomical stations on the surveyed route running about east and west given in the record of measurements for Shikoku were studied, the difference between the geodetic latitude and the astronomical latitude for each station was found to tend to increase gradually as it progressed westward. The comparison of the geodetic materials and the astronomical latitudes of various points on the Nakasen Highway surveyed in 1809 showed a similar tendency. All this proves that the magnetic declination of the area extending over the Nakesendō and Kyūshū about 1809 differed from that of Yedo about 1802 in its direction, and deviated westward.

If the above-mentioned systematic tendency is assumed to have arisen from the magnetic declination and its value is roughly computed by the method of least squares, the value obtained for the southern coast of Shikoku will be $\delta=39'$W., and that for the northern coast $\delta=28'$W. A tentative computation of the data for the western part of the Nakasendō gives $\delta=49'$W. as its result. Of course, the astronomical data availed of include, besides what may be regarded as the ordinary observation error, the systematic error arising from the omission of correction for nutation and aberration. The value thus obtained, therefore, should not be taken as the approximate value of the magnetic declination. However, considering the fact that the three tentative computations from the materials for different times and districts all gave an approximate value of 30′W., perhaps it is very likely that the magnetic declination in the western half of Japan about 1809 was something like this.

This shows that though the magnetic declination in Japan at the time Inō was engaged in surveying was generally small, it was not small enough to be entirely ignored. As for the general feature of its distribution, it seems to have been an eastern declination of 10′–20′ in the Kwantō district, then decreasing gradually in its amount by turning westward in the western districts, a practical zero declination in some zone, and finally a western diclination of 30′ in the Shikoku and Kyūshū districts.

Thus, although the mean error of the magnetic azimuth measured with two or three magnetic theodolites, including the diurnal variation of the magnetic declination was only about 12′, it must be admitted that the value of the error of the measured azimuth as against the true azimuth was often several times as great as this mean error.

Since the magnetic declination generally retains an approximately equal value within a small district, and its influence upon azimuth observation has the nature of constant error, the small-section map drawn from the materials containing the magnetic declination included only a common azimuth error equal to the magnetic declination δ, which had no bad effect upon the relative positions among the stations lying in the area. But when such small-section maps were collected and a larger one was drawn, the magnetic declination had serious effects upon the E.W. direction, and the error gradually accumulated with the increase of the length of the surveyed line, till it exceeded ordinary instrumental and observation errors. (The N.S. direction, being controlled by the results of astronomical observation, caused no appreciable harm.) And if the declination of the western part of Japan is roughly estimated at 30′, the southern section of Kyūshū which was under the most unfavourable conditions, may involve an error of 2′ in longitude as against the Kyōto-Ōsaka region. The Tōhoku district was also under similar conditions; but as the magnetic declination in this district appears to have been comparatively small, it is not likely that the accuracy of measurement was actually affected as much as in the western districts.

3 His Latitude Measurement

Instrumental Errors. The quadrant with a radius of 1·15 m. which Inō chiefly used has suffered various injuries during the long years, and the graduated surface does not preserve the exact conditions of the time when Inō used it. It is impossible to obtain definitely the graduation error of a hundred years ago simply by examining the instrument. To study its accuracy on the whole, however, the graduation error, when the graduated surface was held horizontal, cannot be said to have been kept always below 5″ or 10″, while it is not likely to have reached 1′ even at the worst point. When practically used in a vertical position, the method of fixing the instrument to the stand was so improper that some distortion of the quadrant due to its own weight added a new systematic error to the graduation.

Besides the constant error of the graduation, the chief instrumental errors to be introduced are the index error, the one caused by the eccentricity of the rotation-axis of the index plate, and the one arising from the distortion of the index plate and the telescope. Of all these, Inō himself devised no method of eliminating or correcting the effect of errors, except the index error, from the result of his measurements. The error arising from eccentricity should have reached the maximum value of 30″ even for an eccentricity so small as 0·2 mm., as is proved by our examination of the instrument.

This being the case, when making an absolute measurement of altitude with this quadrant, the instrumental error alone made it almost impossible to expect an accuracy greater than 30″ for the value of a single measurement; and this accuracy tended to decrease as the altitude of the observed object decreased. But when this was used as an instrument for the relative measurement of altitude, it was not difficult to keep the effect of the instrumental error at 10″ or so by skilful handling of the instrument and careful choice of the fixed stars, since such instrumental errors crept in only differentially.

Observation Errors. Of all the errors committed by Inō in measuring the meridian altitude of fixed stars with the quadrant, the sighting error and the reading error of graduation were comparatively slight; the principal error was caused by the imperfect adjustment of the face of the quadrant to the meridian, the instability of the instrument, and the inability to secure a perfect adjustment of the vertical side or take an accurate reading of its deviation. Of these errors, we shall first consider the amount of the error arising from the imperfect coincidence of the surface of the quadrant and the meridian. For setting up the quadrant at a field-station, Inō made it a rule to rely upon the magnetic needle, and moreover, as the device for keeping the surface of the instrument within the meridian was not very delicate, it was not seldom that the azimuth of the quadrant amounted to 1°. Supposing this azimuth to be 1°, the altitude observation of a fixed star culminating southward at an altitude of 55° at a station $\varphi=35°$ would involve an error of 15″, and that of a fixed star culminating northward at an altitude of 55° an error of 43″; and supposing the azimuth to be 30′, the values would be 4″ and 11″ respectively. Again, the adjustment of the vertical side of the quadrant or reading its slight deviation by means of a plumb-line was not to be compared in accuracy with reading the angle by the index plate. Especially when fixed stars culminating in the south and north were observed alternately by turning the quadrant in a short time, reading was no doubt taken hurriedly, without waiting for the plumb-line to settle completely, so that accuracy was further marred and an error 10″ or so always existed. Finally, for the error arising from the instability of the instrument, it increased or decreased according to the degree of care on the part of the observer and his assistant; so it is quite hard to determine the amount, but considering the construction of the instrument and the method of setting it up in the field, even when the utmost care was taken, an error of about 5″ was perhaps inevitable.

Thus, the observation error arising from various causes in an altitude measurement in a field-station may in general fairly be estimated as 25″–30″. And the combined effect of this observation error and the instrumental error previously discussed upon an altitude measurement was probably about 1′ in the case of absolute measurement.

The above is the amount of the error estimated from the extant measuring-instruments, the method of measurement adopted by Inō, and also our own experience. Fortunately there exists a record of measurements from which this error can be computed, and the accuracy of altitude measurement in Yedo more concretely ascertained.

The record is a part of Inō's observation note for compiling a star table, and contains all the apparent altitudes of 34 fixed stars for each night observed at his residence at Kuroye-chō, Fukagawa, Yedo, from Nov. 19th, 1802, to Dec. 22nd of the same year; the number of observations recorded exceeds 300.

Now, these observed apparent altitudes corrected for atmospheric refraction were combined with the declinations of the fixed stars at the time of Inō's observation newly obtained by consulting Newcomb's Catalogue of Fundamental Stars and Auwer's Reduction Table, and the latitude of Kuroye-chō was computed from each observation. Then the instrumental and observation errors were computed from the conditions of the distribution of errors committed by the values of the latitude obtained in the above-mentioned manner; and 30″ was obtained as the approximate value of the instrumental error, and 20″ as that of the error of a single observation. These values are almost equivalent to those previously assumed from the construction of the quadrant.

Moreover, according to all the materials contained in this record 35° 39′ 55″ will be the value for the latitude of Kuroye-chō; but, in order to avoid the effect arising from the graduation error and the distortion of the quadrant, 7 pairs of southern and northern stars out of the 34 with approximate equal altitudes are chosen; and if the latitude is computed exclusively from the observation materials of these 14 stars, the value will be 35° 40′ 8″ and will quite closely approximate the value 35° 40′ 17″ determined by the latest observation materials.

According to what we have considered from every side, the amount of error in the absolute measurement of the altitude of the fixed stars may be inferred as follows:

(1) Though the mean error of observation in a single altitude measurement at Yedo was 20″, it somewhat increased at a field-station—probably to as much as 30″.

(2) The instrumental error (excepting the index error) was probably 30″ in most cases, but some part of the graduation was so deformed as to make the error exceed even 1′.

(3) Consequently, the accuracy of a single altitude measurement, taking both instrumental and observation errors into account, was about 1′ in most cases. For the altitude near 60° the conditions of the instrument proved comparatively favourable and a comparative decrease of the systematic error more or less increased the accuracy. The altitude below 50° increased the systematic error and the neighbourhood of the zenith increased the observation error, because of the unnatural attitude the observer had to take; all this no doubt lessened accuracy to some extent.

(4) A mere increase of the number of the fixed stars or that of their observations would not have secured a remarkable improvement of the result; but if care had been taken in selecting the fixed stars and in eliminating the effect of the instrumental error, and if the number of observations were increased to diminish the observation error, the latitudes of the stations could have been computed as accurately as 10″, as the result of such observations. (Of course, it was important to know definitely the declinations of the fixed stars beforehand. As no star table available for such purpose existed in Inō's days, he was obliged to compile one from his own observations; consequently, he was unable to compute latitude with such accuracy immediately from his own observa-

tion.)

The Accuracy of His Star Table. As already stated, in the star table which Inō compiled were arranged the apparent altitudes at a given period of the fixed stars observed at the fundamental station for several nights or several dozen nights—namely, the mean values of their observed apparent altitudes corrected for index error and precession. Since the values had not been corrected for nutation or aberration, the accuracy tended rather to decrease more or less from that of the mean values themselves of the observed apparent altitudes. When the star table for 1813 now extant is compared with that computed back from the latest star table, the error is generally less than 30″ or 40″; however, the error for some stars which chanced upon a position having a great instrumental error of the quadrant is found to amount more than 1′.

The Accuracy of Latitude Difference. In order to obtain the latitude difference of a field-station, Inō, as previously stated, computed it by comparing the altitude of each of the fixed stars measured at the field-station with that which he had measured with the same instruments at the fundamental station in Yedo and given in his star table. It may be said, therefore, that, to say nothing of the error arising from atmospheric refraction and the effect of change in the positions of the fixed stars, the error of latitude difference computed from the same fixed star was composed of the three following kinds:—an observation error in measuring the altitude of the fixed star at the fundamental station, an observation error in measuring the altitude of the same fixed star at the field-station, and the differential instrumental error arising from a difference of 3 or 4 degrees in position on the quadrant. The mean observation error in a single measurement of altitude at the fundamental station was, as previously stated, 20″ or so; and because the altitude of every star given in the star table had been determined by the results of measuring it generally more than ten times, of total error included in the value, the error considered as observational was $\frac{20''}{\sqrt{10}}$, that is, only 6″ or 7″. The observation error in a single measurement of altitude at a field-station increased slightly from that measured in Yedo, reaching 25″–35″. And the differential instrumental error, arising from a difference of 3 or 4 degrees in position on the quadrant, cannot be estimated definitely to-day; it may be assumed, however, from the records of the measurement of the altitudes of various fixed stars in Yedo, that the error was usually 10″–15″ and its nature appears to have been more or less accidental.

Thus, the mean error caused observationally and instrumentally for the latitude difference computed from a single altitude measurement of a fixed star at a field-station, may be said to be approximately $\sqrt{30^2+7^2+15^2}=34''$. Although partially of a systematic nature, this error was mostly accidental; the mean error of the mean result tended to decrease gradually to some extent, according to the law of errors, with the increase of the number of the fixed stars observed at a field-station. If Inō had been cautious in his preparation of the star table and his computation of latitude, therefore, it would have been possible for him to keep the mean error of the latitude difference computed from the measurement of the altitudes of a large number of fixed stars as low as 10″–15″.

When compiling the star table, or computing latitude difference from the materials of altitude measurement at a field-station, Inō not only disregarded correction for aberration or nutation, but took slight notice of the effect of atmospheric refraction; so that the error arising from these grew to be as great as the observation error or the instrumental.

The error in computation of the latitude difference arising

from omitting correction for annual aberration, though varying according to the date of making an observation and also according to the position of the stars observed, sometimes reached 20″ in the final result, computed from the observation on several stars.

Again, the effect of nutation upon the declination of the fixed stars varies in so long a period as 18 years. As Inō often revised his star table by new data, the date of his observations at the field-stations and that of the compiling of his star table used for it did not differ much. Consequently, the error caused by omitting correction for nutation was, it seems, not generally so conspicuous as that arising from omitting correction for aberration.

Finally, the effect caused upon latitude difference arising from omitting correction for atmospheric refraction creeps in differentially, and the greater the altitude of the fixed star, the less the effect. Inō made it a rule to measure stars with altitudes greater than 40°, and suppose the mean altitude was about 60°, the error could not exceed 10″ at a station with a latitude difference of 6°. So the amount of the error caused by disregarding atmospheric refraction could not be remarkable at a station with a small latitude difference. This, however, always kept the latitude difference too small for a station lying to the north of Yedo and too large for one in the south.

It may be assumed from this that, since the latitude difference Inō himself had computed for his field-stations included such computation errors besides the inevitable observation error and differential instrumental error, the accuracy varied according to the relations between the date of the fundamental measurement and that of the field-station measurement, the appropriateness in selecting the fixed stars on which measurements were made, the amount of the latitude difference, the number of the fixed stars observed, the condition of the weather, and so on; but in most cases the error of the final result did not exceed 30″, though it sometimes reached 1′ under unfavourable conditions. Besides, an examination of the latitude differences of the various field-stations given in the "Kyōwa-ni-nen hokkyoku-kōdo sokuryōki" (The Record of the Latitude Measurement of 1802) which is still preserved, resulted in endorsing this degree of accuracy.

The value 35° 40′ 30″ Inō adopted for the latitude of the fundamental station, or his residence at Kuroye-chō, Fukagawa, Yedo, was greater than the true value by 13″. Therefore, the latitude of each field-station computed by him from the results of astronomical observations, included a constant error of 13″, in addition to the errors already discussed, and had a tendency to make its value generally greater than the true. However, the latitudes of the various stations adopted by Inō in drawing his maps, and registered in the "Enkai-jissoku-roku" (The Record of the Coast-Survey) are those which he obtained first by revising the latitude differences after comparing the results of astronomical observations with those obtained geodetically, and then by adding them to or subtracting them from the latitude of the fundamental station, except those which had been measured before 1801, or those of the special stations which had been used in the determination of the length of one degree of the meridian arc. Hence their accuracy somewhat exceeds that obtained by astronomical observation alone. With a view to illustrating definitely the accuracy of the latitude of each station adopted by him as a final value, the author has freely selected 50 or 60 out of the values of the latitudes for over 1000 stations registered in the "Enkai-jissoku-roku" and compared them with those determined by the most recent observations, with the following results:—

Name of Locality	Record of Coast-Survey		Most Recent Measurement	
	Latitude	Station	Latitude	Station
Fukuye, Hizen	32°41′ 30″	Hamachō	32°42′ 2″	. . . .
Idzuhara, Tsushima	34 12 0	Nakasukachō	34 12 12	Telegraph Office
Nagasaki, Hizen	32 45 0	Rokasuchō	32 44 33	Great Northern Telegraph Co.
Kagoshima, Satsuma	31 36 6	Kurumachō	31 35 40	Telegraph Office
Kumamoto, Higo	32 48 0	Shin-itchōme	32 48 14	Former Governer's Residence
Kokura, Buzen	33 53 30	Sendōchō	33 53 35	Break-water
Akamagaseki, Nagato	33 57 30	Nambuchō	33 57 30	Former Telegraph Office
Ōita, Bungo	33 14 30	Sakurachō	33 14 43	Middle School, (Uyeno)
Nobeoka, Hyūga	32 34 30	Minamichō	32 34 45	Telegraph Office
Yamaguchi, Suō	34 10 30	Nishimonzen-chō	34 10 30	Centre of City
Hamada, Iwami	34 53 30	Shin-machi	34 54 3	Takatayama
Hiroshima, Aki	34 24 0	Sakaimachi	34 23 30	Telegraph Office (Ōtemachi)
Matsuyama, Iyo	33 51 30	Fuchūmachi	33 50 23	Telegraph Office
Matsuye, Idzumo	35 27 30	Suyetsugu-honmachi	35 28 51	Middle School (Kitahori)
Fukuyama, Bingo	34 30 30	Fukatsuchō	34 28 58	Middle School
Kōchi, Tosa	33 34 0	Tanezakimachi	33 33 28	Kōchi Park
Marugame, Sanuki	34 18 0	Tōrichō	34 17 4	Middle School (Rokubanchō)
Okayama, Bizen	34 40 30	Shimo-no-chō	34 39 38	6th High School (Kunitomi)

Tottori, Inaba	35°30′ 0″	Moto-imoji-machi	35°30′ 18″	Normal School (Higashi-machi)
Tokushima, Awa	34 5 0	Shin-uo-machi	34 4 36	Nishi Shin-machi
Himeji, Harima	34 50 30	Tanemachi	34 50 6	Girls' School (Kyōguchi)
Miyadzu, Tango	35 32 0	Uoyamachi	35 32 0	Centre of City
Wakayama, Kii	34 13 30	Jōgai Minato Kubo-machi	34 13 51	Meteorological Station
Ōsaka, Settsu	34 40 0	Tondayachō	34 41 21	Castle-tower
Kyōto, Yamashiro	35 0 30	Shinsenyenchō	35 1 7	Former Meteorological Station (Imperial Garden)
Kushimoto, Kii	33 28 0	....	33 28 4	Nishi-no-oka
Nara, Yamato	34 41 0	Taruichō	34 41 14	Normal School (Noboriōjichō)
Tsuruga, Echizen	35 39 30	Nishihamachō	35 39 14	Kebi Shrine
Kanazawa, Kaga	36 34 30	Owarichō	36 33 53	4th High School (Sengokuchō)
Toba, Shima	34 28 30	Fujinogō	34 28 35	Aioiyama
Nagoya, Owari	35 10 0	Tamayachō	35 9 55	Centre of City (Main Street in front of Railway Station)
Nanao, Noto	37 2 30	Aramachi	37 2 58	Atagoyama
Takayama, Hida	36 8 30	Sanno-chō gochōme	36 8 25	Centre of City
Hamamatsu, Tōtōmi	34 42 30	Hatagochō	34 42 49	Middle School (Motonagori)
Matsumoto, Shinano	36 13 30	Hon-machi	36 14 21	Middle School
Shidzuoka, Suruga	34 58 30	Demmachō	34 58 22	Konyachō-ura
Takata, Echigo	37 7 0	Gofukuchō	37 6 23	Middle School

Kōfu, Kai	35° 39′ 30″	Yanagimachi nichōme	35° 39′ 25″	Centre of City
Shimoda, Idzu	34 40 30	Minato	34 40 20	Takegahama
Niigata, Echigo	37 55 30	Furusannochō	37 56 6	Shin-hiyoriyama
Yedo, Musashi	35 40 30	Kuroye-chō, Fukagawa	35 40 17	Kuroye-chō, Fukagawa
Wakamatsu, Mutsu (Iwashiro)	37 30 0	Nanukamachi	37 29 33	Middle School (Sakaye-machi)
Utsunomiya, Shimodzuke	36 33 30	Ikegamichō	36 34 9	Hachiman-yama
Tsuchizaki, Dewa (Ugo)	39 45 0	....	39 44 56	County Office
Mimmaya, Mutsu	41 11 50	....	41 11 35	Centre of City
Shinjō, Dewa (Uzen)	38 45 30	Minami-hon-machi	38 46 2	Middle School
Chōshi, Shimōsa	35 43 0	Iinuma	35 43 31	Meteorological Station
Sendai, Mutsu (Rikuzen)	38 16 0	Kokubunchō	38 14 47	Atagoyama
Midzusawa, Mutsu (Rikuchū)	39 9 0	....	39 8 4	Latitude Observatory
Morioka, Mutsu (Rikuchū)	39 43 0	Kokuchō	39 41 55	Centre of City
Noheji, Mutsu	40 52 30	....	40 51 50	Centre of City
Miyako, Mutsu (Rikuchū)	39 39 0	Kuwagasaki Minato	39 38 20	Mouth of Miyako River
Hakodate, Yezo (Oshima)	41 47 0	Jizō-machi	41 46 7	Naval Coaling-Station
Kushiro, Yezo (Kushiro)	42 58 0	....	42 58 8	Uramichō

Since the new and old stations are more or less different in their positions, their values given in this table should not be strictly compared with one another. When we examine the values of the latitudes adopted by Inō, taking this difference of position into consideration, their accuracy is found to increase more or less from that determined only by the results of astronomical observation, their error being kept below 30″ and seldom exceeding 1′ even in extreme cases. In other words, the latitudes registered in the "Enkai-jissoku-roku" with 30″ as their least unit may be said generally to give correct values within the unit.

4 His Adopted Length of One Degree of the Meridian Arc

28·2 *ri,* adopted by Inō for the value of the length of one meridian degree, was computed, as stated previously, making use of the materials obtained chiefly in the survey of the north-eastern part of Japan. It might also be called the mean length of one degree of the meridian arc between 35° and 41°. In order to examine the accuracy of this length, the value 28·2 *ri* may be converted to 110·85 km. by making use of the equation: Inō's 1 *shaku* = 0·3033 m. And the mean length of one degree of the meridian arc between 35° and 41° computed from Bessel's constants is 110·98 km. The difference between these two values—0·13 km.—should be regarded as the error of Inō's length of one meridian degree. It amounts to about a thousandth of the whole length.

In computing the length of one meridian degree from the longest surveyed route, extending over about five degrees and a half, between Yedo and Mimmaya, an error of this amount was to be introduced even when no error was included in the other materials, if only an error of 20″ existed in the value of the latitude difference between the terminal stations. It is no wonder, that Inō's length of one meridian degree which was computed from the materials including various errors, should have contained an error of such an amount as 1/1000; it is rather wonderful that it should have agreed so closely. Moreover, that the value obtained was smaller than the true one, in spite of the fact that more than one of the elements introducing the error were of a systematic nature, likely to make the computed length of one meridian degree exceed the true value, proved of unexpected advantage afterwards when this value was used in adjusting the survey of the south-western districts of Japan. For the south-western districts, whose topographical survey Inō made most minutely in the latter half of his surveying of the whole of Japan, lay between 31° and 36° in latitude, and the true mean length of one meridian degree in these districts was 110·90 km., differing only 0·05 km. or 1/2000 of the whole length from the length of one meridian degree as adopted by Inō.

Then, if the earth were supposed to be a sphere with 28·2 *ri* for one degree of the great circle, according to Inō, the length of one second of the parallel of latitude for 30°, 35°, and 41° would be by the metric system, 26·666 m., 25·222 m., and 23·238 m., respectively, while the values computed for the spheroidal earth from Bessel's constants would be 26·799 m., 25·355 m., and 23·368 m., respectively; and the error would be about 1/200 of the whole length, by no means a small amount.

If the longitudes and latitudes of the various stations were determined geodetically by the length of one meridian degree as adopted by Inō and by assuming the earth to be a sphere, errors of such extent would be introduced, even when no error was present in the survey-materials. To make this point definitely clear, the values of the errors to be introduced in the geodetic computations of the longitudes and latitudes of the principal and the distant stations lying within the districts Inō surveyed (Kyōto being taken as the origin of computation) are given in the following table.

Station	Position of Station (Approximate)		Error of Latitude	Error of Longitude
	φ	λ	φ(Bessel) − φ(Inō)	λ(Bessel) − λ(Inō)
Sendai	38°15′N	5° 8′E	−12″	−1′·6
Mimmaya	41 12 ,,	4 44 ,,	−28	−1·5
Yedo	35 40 ,,	4 3 ,,	− 2	−1·3
Niigata	37 56 ,,	3 18 ,,	−10	−1·1
Toba	34 29 ,,	1 5 ,,	+ 1	−0·3
Kōchi	33 34 ,,	2 13 W	+ 3	+0·7
Hiroshima	34 24 ,,	3 17 ,,	+ 1	+1·0
Akamagaseki	33 58 ,,	4 48 ,,	+ 2	+1·5
Kagoshima	31 36 ,,	5 11 ,,	+ 6	+1·6
Nagasaki	32 45 ,,	5 52 ,,	+ 4	+1·8

It will be noticed that the error in latitude is comparatively small, remaining mostly below 10″, and not reaching 30″ under most unfavourable conditions, while the error in longitude is considerably greater, reaching almost 2′ at the extreme stations. And this smallness of the error in latitude was no doubt one of the causes that enabled Inō to fulfil his purpose of increasing accuracy by comparing and revising the latitudes of the various stations he determined astronomically and those he determined geodetically.

5 His Longitude Measurement

It has been remarked elsewhere that Inō's effort to adjust his geodetic error by determining longitude by astronomical observation resulted almost in a total failure. Hence it is not very important to inquire carefully into the error introduced into the longitude astronomically obtained. Nevertheless, a consideration as to wherein he failed may not be altogether useless.

Let us consider the nature of the eclipses of the sun, the moon, and Jupiter's satellites, which Inō made use of in his longitude measurement. Since these phenomena all progressed slowly, it was quite difficult to measure accurately the time when the special aspects such as the first and the last contacts and a maximum obscuration presented themselves; even in the case of a solar eclipse which may be rather accurately observed, it is feared that Inō even then committed an error amounting to 10^s in his observation of the first and the last contacts; in the case of the eclipses of the moon and of Jupiter's satellites, his observation error was probably increased, often to an amount several times as great as this. And though Inō generally employed three or four men in the observation of an eclipse and, consequently, the error for the mean value was probably decreased according to the law of errors, it may be considered that the amount still reached 5^s in the case of the solar eclipse and a value several times as great as this in the case of the eclipse of the moon or Jupiter's satellites.

Again, there exist to-day very few of the materials for investigating the accuracy or the state of variation in the daily rate of the pendulum-clock used by Inō in indicating the local time of a station; only a record giving the daily numbers of the oscillations of the clock deduced from his observation of some fixed stars for successive nights from March 15th to 20th, 1802, exists.

According to the record, the pendulum-clock used by Inō seems to have been in a rather satisfactory condition during this period; the number of its oscillations was about 59504 a day, the daily variation never exceeding two oscillations, or about 3 seconds at the most. But the iron wire used for the stem of its pendulum caused a periodic irregularity to the oscillation, according to the change of temperature, and consequently the actual accuracy more or less fell below the mean accuracy.

In determining local time with the transit-instrument, Inō made his observation and computation, assuming that his instrument had been adequately adjusted. It was quite frequent, however, that the instrument was not correctly adjusted, and consequently a considerable error was introduced into his determination of noon. For instance, by making use of the materials he obtained by measuring the fixed stars on the night of March 15th, 17th, and 20th, 1802, and the right ascensions of the fixed stars newly computed backwards from the latest star table, the author has obtained $5'-6'$ for the approximate value of the azimuth of the transit-instrument at the time Inō made the observations. This was an error introduced in adjusting the instrument. Thus, if the declination of the sun at the time had been $0°$, an error of 12^s-14^s would have arisen in determining local noon. And since these observations were made at his own residence in connection with an observation of a lunar eclipse, it cannot be denied that Inō was particularly cautious in his adjustment of the instrument. If he made an error of this amount through the inadequacy of adjusting his instrument at his base station, the error due to the same cause at a field-station, on account of hasty preparation, may be presumed to have been generally not smaller than this.

This shows that though Inō's failure in determining longitude by astronomical observation was partly due to the unsuitableness of the nature of the phenomena for accurate observation, the large error introduced in his determining local time may be regarded as another chief cause.

As stated above, his astronomical determination of longitude resulted in failure, and the longitude of each station had to be determined entirely from geodetic data. If Inō had adopted an adequate method in the arrangement of survey-materials and in the construction of maps, the accuracy could have been tested concretely by comparing the longitude of each station on the map with the value of the latest measurement. As already stated, however, Inō made no reasonable computation in determining the E.W. position of each station, nor did he take sufficient care in constructing the complete map; on the contrary, his methods often spoiled the accuracy of his measurements. It is impossible to examine the actual accuracy of his geodetic longitudes by such a method. Therefore, we have to go back to examine the accuracy of the fundamental materials, such as distance and azimuth measurements, which Inō made use of in computing his geodetic longitudes. We have also to consider the influence of the magnetic declination, and consult the extant fragmentary record of his computations, and thereby to infer generally the actual accuracy which could have been expected for his geodetic longitudes.

As for the longitude difference between two stations connected by several surveyed routes, such as the longitude difference between Yedo and Kyōto, or that between Kyōto and Akamagaseki, it is probable that the error, excepting that arising from ignoring the ellipticity of the earth, did not exceed 1/300 or 1/400 of the actual value. Consequently, the longitude of each station lying in these districts was fairly accurately determined geodetically and it may be allowable to presume that the accuracy was about a half of that of the latitudes adjusted by his astronomical materials. On the contrary, along such surveyed routes as the one between Akamagaseki and Kagoshima, or the one between Yedo and Mimmaya, running nearly north and south and through the well-known volcanic districts, the determination of the E. W. positions of the various stations lying in the districts was sometimes influenced considerably by the ordinary magnetic declination and the local magnetic disturbance. To make matters worse, the great distance from Kyōto, the standard point of his longitude computation, increased the error of measurement accordingly and the accuracy of the longitudes geodetically determined for these stations was inevitably halved in comparison with that for the stations in the Tōkai or the Chūgoku district. In other words, an adequate adjustment of Inō's geodetic materials, even ignoring the influence of the magnetic declination, could have determined the longitudes of the stations in the Tōkai or the Chūkogu district, which were measured from Kyōto, with an error less than 1′ in most cases, and those of the stations, in the Tōhoku district or the southern part of the Kyūshū district, with an error of about 2′ or a little more; and, if it had been possible to eliminate the influence of the magnetic declination, even the error for the stations at the remotest frontiers could have been kept generally below 1′. The geodetic materials of the Yezo district alone, chiefly obtained single-handed by Rinzō Mamiya, were not very accurate, the longitudes geodetically computed for the stations in this district including a rather considerable error, which seems to have reached 10′ in extreme cases.

6 His Maps

If Inō had adopted an adequate method in constructing maps by the materials obtained in his survey, the accuracy of his maps would not have been lower than that of his survey-materials—it could have been rather improved by the mutual adjustment and revision of the various materials. As already stated, however, Inō quite frequently failed to adopt proper measures in his map-construction, and, therefore, the accuracy of the maps can not be inferred directly from the accuracy of the survey-materials. Besides, it was not uniform, but varied according to the kind of map. The accuracy of each kind of map will be separately considered.

The Accuracy of the Small-Section Map. As for the small-section map which Inō projected as his first step in map-construction for each tract of land 5–10 km. square, the error introduced through his assuming the surface of the earth to be a plane was so much smaller than the error of measurement that it might well be ignored. Neither is his method of the mutual adjustment of the survey-materials open to criticism. The accuracy of the map of this kind may be said to be greater than that of each of the survey-materials.

The Accuracy of the "Combined" Map. On the so-called "combined" map, constructed by combining small-section maps for an area 40–50 km. square, the error introduced by assuming the surface of the earth as a plane increased with the extension of the area, rivalling the error of measurement. Besides, his unsuitable and temporizing method of adjusting it tended to decrease the accuracy of the map somewhat below that of the survey-materials. However, since the error caused by the unsuitableness of projection never exceeded 5–10 mm. on the 1 : 36000 reducing-scale, even in extreme cases, it practically caused little inconvenience in obtaining the relative positions of the stations given on the map. In measuring azimuth and distance on the map, the decrease of accuracy through lack of care in map-construction need not be seriously considered, in view of the inaccuracy inevitably caused by the deformation of the drawing-paper.

The "combined" map only formed a draft for constructing the district or the complete map; the map made public was the large-scale map which had been developed from the "combined" maps. As has been discussed in the paragraphs on map-construction, the large-scale map was made in the following way; first some "combined" maps were reduced to the scale of 1 : 216000 and then, after some correction, were collected into a district map, which was in turn reproduced to the scale of 1 : 36000, and then cut into maps of the desired size. Therefore, if the method of adjustment and projection adopted in his construction of the district map had been reasonable, the accuracy of the large-scale map would have more or less exceeded that of the "combined" map. In making a district map from the "combined" map, Inō adopted Sanson and Flamsteed's system of projection which considerably lost the conformal relation on the borders of the map; but still his wish as far as possible not to lose such relation and the lack of caution in his method of construction and correction, introduced some projection error. Consequently, the large-scale map which should not have retained the conformal relation from its very nature, did retain it to some extent in practice, and became a strange thing after all. As a result, not only was the accuracy of the large-scale map inferior to that of the "combined" map, but the latter because of its genuineness, proved of more value as a map than the former.

The Accuracy of the District Maps. In collecting "combined" maps into a district map, Inō generally adjusted them at his own discretion, and often smoothed over discrepancies in a temporizing way; consequently, in spite of the presence of rather considerable slant intersections of the meridians and parallels towards the borders of the district map, it was quite a singular spectacle that the topographical features projected in such parts of the map were frequently similar to those of the "combined" map, retaining much of the conformal relation. Besides, in drawing the meridians and parallels he added a new kind of drawing error, because of ignoring the ellipticity of the earth. Thus, the accuracy of the map (that is, the accuracy of the positions of the various stations referring to the meridians and parallels of the map) was not uniform, but varied according to circumstances; yet, the accuracy of longitude and azimuth was generally somewhat lower than that to be reasonablly expected from the accuracy of the survey-materials. Only latitude, being controlled by the results of astronomical observation, was free from the introduction of any drawing error, and retained its own proper accuracy.

In order to illustrate the amount of the error of the longitude of each station on the district maps now extant, let us compare the longitudes of the various stations measured, by referring to the meridians and parallels on the district map (where no meridians were drawn, the author, according to the manner shown in other maps, has filled in the meridians so that the one passing the central part of the map might intersect the parallels perpendicularly) and the results based on the latest measurements. Of all the district maps, that of the Tōkai Highway and the coast of the Chūgoku district has the smallest error of longitude. Although this district map extended over 4° both eastwards and westwards from Kyōto, the centre, the positions of the various stations are projected with comparative accuracy; the error of longitude is almost always kept below 1′, and the errors for the various stations in the Sanyō district are especially small. The accuracy of the district maps of Shikoku and of Kyūshū is slightly lower; at a station with an approximate longitude difference of 1° 30′, the error of longitude sometimes exceeds 2′. The accuracy of the district map of the eastern coast of the Main Island is so inferior that the error of longitude for some station in the north-eastern part amounts as much as 7′.

This aggravation of the error of longitude in the district maps is chiefly due to Inō's faulty projection of the maps; but it is evident that his ignoring the magnetic declination and the ellipticity of the earth certainly was a secondary reason. The remarkable smallness of the error in the map of the Chūgoku district was due to the fact that the special features of the surveyed lines composing the district map were such that the error caused by ignoring the magnetic declination and that due to the ellipticity of the earth, for the most part, cancelled each other.

Let us now consider the accuracy of the azimuth indicated by the azimuth lines drawn among the various points on the district maps. To begin with, the azimuth lines on the district maps drawn by Sanson and Flamsteed's projection should not be generally represented by straight lines. Notwithstanding this, Inō always represented them with straight lines, without paying any serious attention to this point and, therefore, the azimuths shown on the district maps included a sort of drawing error. The amount of this error was not uniform, but varied according to the positions, distances, and directions of the stations, and frequently exceeded the error of the azimuth measurement, reaching in some cases an amount several times as great as the latter. For this reason, the numerical values given along the azimuth lines on the maps are not the exact values of the azimuth obtained by his measurements, but the values rounded to the unit of 1°·5, or one-twentieth of 30°.

The Accuracy of the Complete Map of Japan. As already explained, on the complete map of Japan, finished in 1821, a ridiculous blunder was committed, rendering the meridians drawn on the map meaningless. It is true that fortunately the latitudes of the stations on the map did not suffer evil effect from this, but the longitudes fell in so serious an error that an inspection of the positions only on the map would necessarily lead one into a terrible misconception as to the accuracy of Inō's measurements. The author has for reference measured the longitudes of some stations on the complete map of 1821 according to the meridians marked on it, and compared them with the results of the latest longitude measurements. The considerable differences between the two will be given in the following table.

It should be remarked that the longitude of the site of the Calendaric Office at Nishi-Sanjō-dai, Kyōto, (the origin of longitude compution) has been taken to be $9^h\ 2^m\ 58^s{\cdot}4 = 135°44'\ 36''$ E. of Greenwich by the various materials of the latest measurements; and the longitudes of the other stations obtained by the latest measurements have been reduced to that counted from the Calendaric Office. And in order to measure the longitudes of the stations on the complete map according to the meridians on it, the middle-scale map in the possession of Viscount Ōkōchi, which is preserved in the best condition of all the maps of the kind, was used; in spite of this, an inaccuracy $0'{\cdot}2$ or $0'{\cdot}3$, due to the expansion and contraction of the drawing-paper, was perhaps inevitable.

Name of Locality		Complete Map		Latest Measurement		Difference	Latitude
		Longitude	Station	Longitude	Station		
	Kyōto	0° 0′·0	Site of Calendaric Office, Nishi-Sanjō-dai	0° 0′·0	Site of Calendaric Office, Nishi-Sanjō-dai	—	35° 0′·8
Tōkai District	Wakayama	0 34·6 W	Centre	0 34·9 W	Meteorological Station	− 0′·3	34 13·8
	Toba	1 4·4 E	Centre	1 5·4 E	Aioiyama	+ 1·0	34 28·6
	Nagoya	1 9·6 E	Centre	1 9·6 E	Centre	0·0	35 10·0
	Shidzuoka	2 38·0 E	Centre	2 38·9 E	Konyachō-ura	+ 0·9	35 58·4
	Shimoda	3 9·9 E	Centre	3 12·4 E	Takegahama	+ 2·5	34 40·3
	Niijima	3 28·4 E	Miyadzukayama	3 31·6 E	Miyadzukayama	+ 3·2	34 23·6
	Ōshima	3 39·2 E	Habu Bay	3 41·8 E	Habu Bay	+ 2·6	34 41·2
	Miyakejima	3 42·8 E	Oyama	3 47·3 E	Oyama	+ 4·5	34 5·5
	Mikurajima	3 46·7 E	Centre	3 51·7 E	Centre	+ 5·0	33 52·5
	Yokosuka	3 52·8 E	Centre	3 55·4 E	. . . . . .	+ 2·6	35 17·1
	Hachijōjima	3 51·4 E	Ōga-gō	4 1·6 E	Ōga-gō	+ 10·2	33 5·8
	Yedo	4 3·7 E	Kuroye-chō	4 3·3 E	Kuroye-chō	− 0·4	35 40·3
	Chōshi	5 5·5 E	Western End	5 5·0 E	Western End	− 0·5	35 43·7

Hokuriku District	Tsuruga	0°20′.9E	Centre	0°19′·6E	Kebi Shrine	−1′·3	35°39′·2
	Nanao	1 16·5 E	Centre	1 12·6 E	Atagoyama	−3·9	37 2·5
	Amaze	3 0·7 E	Centre	2 56·5 E	Myōfukuji	−4·2	37 32·5
	Niigata	3 24·2 E	Centre	3 17·8 E	Shin-Hiyoriyama	−6·4	37 56·1
Tōhoku District	Tsuchizaki	4 33·1 E	Centre	4 19·4 E	County Office	−13·7	39 44·9
	Aomori	5 20·2 E	Centre	5 0·4 E	Near Telegraph Office	−19·8	40 49·9
	Sendai	5 15·6 E	Centre	5 8·1 E	Atagoyama	−7·5	38 14·8
	Midzusawa	5 33·0 E	Centre	5 23·4 E	Latitude Observatory	−9·6	39 8·1
	Miyako	6 27·4 E	River Mouth	6 13·8 E	River Mouth	−13·6	39 38·3
Yezo District	Hakodate	5 23·0 E	......	4 58·6 E	Naval Coaling-Station	−24·4	41 46·1
	Otaru	5 49·8 E	......	5 15·9 E	Suitengūyama	−33·9	43 11·7
	Wakkanai	6 49·0 E	......	5 55·9 E	Former County Office	−53·1	45 25·0
	Urakawa	7 30·2 E	......	7 1·9 E	Site of Light House	−28·3	42 9·2
	Abashiri	9 14·6 E	......	8 31·9 E	Nakatōrichō	**−42·7**	44 1·2
	Kushiro	9 21·4 E	......	8 37·9 E	Uramichō	−43·5	42 58·1
	Nemuro	10 36·6 E	......	9 50·6 E	Telegraph Office	−46·0	43 20·0

Chūgoku District	Ōsaka	0°15′·0W	Tondayachō	0°15′·1W	Tondayachō	−0′·1	34°40′·6
	Miyadzu	0 31·7 W	Centre	0 32·9 W	Centre	−1·2	35 31·9
	Himeji	1 2·5 W	Centre	1 3·0 W	Centre	−0·5	34 49·8
	Tottori	1 30·0 W	Centre	1 30·7 W	Centre	−0·7	35 29·8
	Okayama	1 48·9 W	Centre	1 48·7 W	Centre	+0·2	34 39·5
	Matsuye	2 39·8 W	Centre	2 41·2 W	Centre	−1·4	35 27·7
	Hiroshima	3 16·8 W	Centre	3 17·6 W	Centre	−0·8	34 23·5
	Hamada	3 38·2 W	Centre	3 39·7 W	Centre	−1·5	34 53·6
	Hagi	4 19·9 W	Centre	4 20·4 W	Centre	−0·5	34 24·5
	Akamagaseki	4 47·0 W	Centre	4 47·9 W	Former Telegraph Office	−0·9	33 57·5
Shikoku District	Okazaki-mura, Awa	1 6·0 W	Centre	1 7·3 W	Centre	−1·3	34 10·7
	Takamatsu	1 41·1 W	Centre	1 41·4 W	Centre	−0·3	34 20·4
	Kōchi	2 10·5 W	Centre	2 12·6 W	Park	−2·1	33 33·5
	Matsuyama	2 57·5 W	Centre	2 58·9 W	Telegraph Office	−1·4	33 50·4
	Yawatahama	3 17·5 W	Centre	3 19·1 W	Daihōji	−1·6	33 27·6

Kyūshū District	Tsurumizaki, Bungo	3° 36′·3 W	......	3° 40′·4 W	......	− 4′·1	32° 56′·3
	Nobeoka	4 0·8 W	Centre	4 4·2 W	Telegraph Office	− 3·4	32 34·7
	Kokura	4 50·5 W	Centre	4 51·8 W	Centre	− 1·3	33 53·5
	Kumamoto	4 59·5 W	Centre	5 2·4 W	Former Governor's Residence	− 2·9	32 48·2
	Kagoshima	5 5·0 W	Centre	5 11·2 W	Telegraph Office	− 6·2	31 35·7
	Misumi	5 13·0 W	Centre	5 17·4 W	Telegraph Office	− 4·4	32 37·5
	Nagasaki	5 48·7 W	Beach	5 52·2 W	Great Northern Telegraph Co.	− 3·5	32 44·5
Western Islands	Shimo-Koshikijima	5 55·8 W	Aose-mura	6 0·9 W	Aose-mura	− 5·1	31 40·3
	Gōura, Iki	6 1·8 W	Centre	6 3·4 W	Yokoyama-no-tsuji	− 1·6	33 44·8
	Idzuhara, Tsushima	6 26·5 W	Centre	6 27·2 W	Telegraph Office	− 0·7	34 12·2
	Fukuye, Gotō	6 49·4 W	Centre	6 53·9 W	......	− 4·5	32 42·0

The state of the distribution of errors in this table may appear at the first glance so chaotic as to admit of no systematization. However, as has been repeatedly stated, these errors have been introduced chiefly through collecting and projecting the district maps to construct the complete map of Japan, and if the table is studied carefully, a definite rule may be discovered in the distribution of the errors. The values of the errors may be considerably decreased by suitably shifting the meridians according to the cause of the errors introduced.

In other words, for the Tōhoku district, the Tōkai district, and the eastern part of the Hokuriku district, simply redraw the meridians, so that the one passing through Yedo may intersect the parallels perpendicularly and measure the longitudes of the various stations according to the new meridians. The errors will at once decrease as follows:—

Tsuchizaki	+0′·1	Aomori	−1′·8	Sendai	+1′·0	Midzusawa	+2′·0
Miyako	−0·2	Hakodate	−2·6	Amaze	+1·6	Niigata	+0·9
Shidzuoka	−1·3	Shimoda	−0·5	Niijima	−0·4	Ōshima	−0·3
Miyake-jima	−0·1	Mikura-jima	−0·2	Yokosuka	+1·4?	Hachijō-jima	+2·8
Yedo	−0·4	Chōshi	−0·4				

Again for the Yezo district, redraw the meridians so that the one passing the central part of the district may intersect the parallels perpendicularly and measure the longitude referring to the redrawn meridians, making connection with Kyōto through the stations at the northern end of the Main Island. Then the errors at any station will be reduced to the amount of 3′–5′ as shown below, except the stations in the south-eastern part where the survey was very imperfect.

Hakodate	−2′·6	Otaru	−1′·9	Wakkanai	−3′·9	Urakawa	−3′·8
Abashiri	−5·0	Kushiro	−13·2	Nemuro	−13·1		

For the Kyūshū district, redraw the meridians, such as the one passing through Kokura situated in the central part of the district to intersect the parallels perpendicularly and measure the longitudes of the various stations from Kyōto according to the redrawn meridians and in connection with the stations in the neighbourhood of Akagamaseki. Then the errors will decrease as follows, and the results will be almost satisfactory, excepting for the stations whose positions have been determined by the method of intersection.

Tsurumi-zaki	−0′·8	Nobeoka	+0′·9	Kokura	−1′·3	Kumamoto	+0′·8
Kagoshima	+1·1	Misumi	−0·3	Nagasaki	+0·2	Aose	+2·1
Gōura	−1·1	Idzuhara	−1·8	Fukuye	−0·5		

But the redrawing of meridians given here is merely a general correction for the drawing error introduced in compiling the complete map of Japan. If the error committed through lack of care in projection may be defined for each locality and correction made for it, it is obvious that the error of the longitude of each station will be still decreased. Hence it is profoundly to be regretted that Inō and his pupils failed to use care in their map-construction.

And what is more regrettable than anything else is that those who looked at the complete map of Japan never suspected the existence of such blunders in its projection, but understood the meridians and parallels on the map to have been so drawn as to be in concordance with the actual projection, thus falling into an unforeseen (or rather reasonable) misconception. Siebold himself was one of those who believed this; as he published in

Europe the complete map of Japan he had secretly taken away with him, he changed the system of drawing meridians and parallels to Mercator's projection, but adopted the positions of the various stations which had been obtained according to the meridians and parallels drawn on the complete map of Japan; he was unable to detect the drawing error which had been introduced in compiling the complete map. Not only Siebold, but even Krusenstern, who professed to be an authority in geography at that time, fell into the same error. As Krusenstern examined the complete map of Japan which Siebold had taken away, he pointed out the presence of some difference between the longitudes of the eastern stations of Yezo given in the map and those which had been measured by Golownin, but it never occurred to him that this difference had arisen chiefly through drawing error. He admitted that the longitudes and latitudes given on the map for the other districts of the country were very accurate. The most ridiculous thing is the fact that, despite the presence of a drawing error of about 4′ in the longitude of Nagasaki from Kyōto on the complete map of Japan, the longitude of Nagasaki from Greenwich contained apparently no error, because an error of about the same amount existed in the value of the longitude of Kyōto from Greenwich which Siebold took away, (according to the map Siebold brought home, the longitude of Kyōto was given as 135° 40′ East of Greenwich) and these errors cancelled each other. Therefore, Krusenstern, inferring from the coincidence of the longitude of Nagasaki for Greenwich obtained from the map with the value he himself had obtained by the method of the lunar distance at Nagasaki some years before, made the following remarks in praise of the great skill of the Japanese astronomers in the measurement of longitude:—

"Dass die Länge der Stadt Kioto, durch welche der erste Meridian gezogen ist, sehr gut von den japanischen Astronomen bestimmt ist, beweist die Länge von Nagasaki, die auf der Japanischen Karte 5° 48′ im Westen von Kioto liegt, d. h. 129° 52′ O. Greenw. bis auf eine Minute die nämliche, welche Dr. Horner und ich aus mehr als tausend beobachteten Mondabstanden berechnet haben."

7 Conclusion

A summary of what has been discussed so far relative to the accuracy of Inō's measurements and maps and the author's own opinion concerning it are given below:—

In the light of the present geodesy, the instruments and methods adopted by Inō, of course, may not be commended for precision and accuracy; considering the early period of more than 120 years ago, when the new sciences had just begun to bud in Japan, the precision of his measuring-instruments is really wonderful. And also that his method of measurement which was continually improved with experience in field-work bears the distinct marks of his incessant strenuous efforts and that he devised various means for avoiding unknown mistakes that might be introduced into surveying, should really impress us with the thoroughness of his caution. But if he had examined the accuracy of his measuring-instruments, investigated the mutual relations of the materials obtained by his measurements, and thereby defined the accuracy that might be reasonably expected from the results of measurement, and engaged himself in the work of surveying, keeping this always in mind, it would have been possible for him to have saved useless labour,—for instance, by discarding the observation of Jupiter's satellites, on one hand, while on the other hand, he could have obtained more or less greater accuracy,—for instance, by using care in the combination of the fixed stars when measuring their altitudes, and also by detecting the magnetic declinations by comparing the magnetic and astronomical azimuths at various stations.

Although there was comparatively little to criticize in Inō's surveying operations in the field, as for his reduction of the survey-materials and his map-construction, performed indoors, there was very much to be desired even in the light of the scholarship of the Japanese astronomers of his day. The chief points are (1) that, in reducing the observed apparent altitudes of the fixed stars, he failed to correct for atmospheric refraction which was well-known at the time and also for nutation or aberration, and thus spoiled the accuracy of astronomical observation; (2) that he added up the so-called E.W. components thoughtlessly and, applying this to extensive areas, lost its proper significance; (3) that he ignored the ellipticity of the earth in his computations of longitude and latitude; (4) that he had neither settled conviction in projecting a map, nor sufficient realization of the nature of a map projected on a plane, and fell into an ambiguous method of adjustments in collecting small-section maps into a larger; and (5) that in compiling the district maps into the complete map he allowed the serious blunder to creep in.

If Inō's field-notes were all preserved in perfect condition, or the small-section maps together with the "Santo-hōiki" (The Record of the Azimuths of Mountains and Islands) and the "Hokkyoku-kōdo-sokuryōki" (The Record of Latitude Measurements) at least were all preserved, we might be able to adjust them reasonably so as to construct a map far more accurate than Inō's. Now that most of the survey-materials have been lost, his great task can only be imagined by the district maps, which contain comparatively little error of computation and inaccuracy of map-construction, or by the complete map of Japan, giving it some suitable general correction in order to remove the gross blunder committed in its compilation.

CHAPTER VII

THE LABOUR AND COST OF HIS SURVEYS

1 Labour

The number of members in Inō's surveying party and their servants was five or six in the earlier periods of his surveying, and fifteen to eighteen at the later, and the number of men and horses hired on the way for helping in the surveying and for carrying measuring-instruments and luggage was seven or eight labourers and two or three horses in the earlier periods, and about ten labourers and six or seven horses later. It is a fact, however, that, besides this, some assistance was rendered by the local officials and common people along the routes. Especially after 1805 Inō was engaged in surveying as a regular official of the Shogunate Government; wherever he travelled he not only levied and used a large number of labourers as carriers of his measuring-instruments and assistants for surveying, but also requested the village officials along the route to submit to him for reference all geographical materials of the route, the old village maps, and the sketch-maps newly drawn at his request. It was his principle to carry out an accurate survey of every coast-line at any cost, in defiance of rough roads, or even where there were none. Thus, it cost the village people tremendous labour to repair the roads and otherwise to prepare for Inō's survey-party. In a number of cases, it is recorded that a village furnished over a hundred labourers.

Though the number of labourers and horses employed in connection with Inō's survey was thus great—several times as great as he was authorized to employ officially—he always put some member of the party in charge of any piece of work upon which the accuracy of his survey depended; even the stretching of the measuring-rope was performed under strict supervision, never being left to outsiders. The assistants for the survey provided by the local authorities were used chiefly in forwarding and transporting measuring-ropes, magnetic theodolites, etc., in erecting measuring-poles and sign-posts for distant sighting, and in other miscellaneous works. Therefore, even when the members worked in three parties, the auxiliaries employed on the immediate work must have been from fifteen to twenty in number.

Inō made it a rule to set to work early in the morning and to quit work also early in the afternoon; in ordinary cases, those in charge of surveying commenced work soon after sunrise and stopped work in the afternoon, varying according to the length of the day—at about 2 p.m. at the earliest, but about 4 p.m. on ordinary days, after which they rested at the inn.

For those in charge of arranging the field-notes and maps all day long, and for those partaking in astronomical observation at night, there was perfect discipline to attend to the official routine. The men furnished by the local authorities to help the party worked in preparing and in straightening things up, sitting up later than the members of the party; especially were the village officials in charge and the reception-committee anxious that the surveying in their village should proceed readily and smoothly, and so they exhausted themselves running to and fro from dawn to dusk, frequently working all night long when any trouble arose.

2 Cost

It may not be entirely useless to state roughly how much it cost Inō to complete his surveys.

Since it was through his petitioning the Shogunate Government that he commenced his surveys, the allowance he received from the Shogunate Government during the earlier periods was extremely small, and as the balance was his own responsibility, he tried to keep the expenses as low as possible. The money expended in 1800–1803, the period for surveying the eastern half of Japan, was 203 *ryō* from the Shogunate Government, and about 140 *ryō,* from his own purse, that is only about 340 *ryō* all told. On the other hand, when he surveyed the western half of the country, the expenses mounted up, because it was undertaken as an enterprise of the Shogunate Government. According to some documents still preserved, the amount paid by the Shogunate Government as travelling expenses of the members of the survey-party, expenses for drawing maps and stationery from 1805 when the survey of the western half of Japan was undertaken until 1821, when the complete map of Japan was finished, must exceed 5500 *ryō*. In other words, the sum expended for the entire survey must have reached 6000 *ryō*. But this is only the amount paid by the Shogunate Government and Inō himself. Besides this, Inō used a large number of labourers free of charge, in the name of the Shogunate Government; this labour alone may be estimated as exceeding 1000 *ryō*; and also, if the reception-expenses spent by the villagers for receiving the members of the surveying party are considered, the entire amount expended for his surveys may double the above-mentioned total of 6000 *ryō* or even go beyond that figure. As one *ryō* in Inō's days corresponds to about ¥12 of to-day, the entire amount, supposing it was estimated at twice 6000 *ryō* or 12000 *ryō*, would reach about ¥150000 according to our present day standard.

CHAPTER VIII

HIS RELICS

1 Measuring-Instruments

The measuring-instruments used by Inō have already been described. Those extant are as follows: a quadrant of a radius of 1·8 m., a quadrant of a radius of 1·15 m., a pendulum-clock, an eclipsemeter, fragments of simpler eclipsemeters, two telescopes, two 1 *shaku* scales, a perambulator, several small magnetic theodolites, two semicircular magnetic theodolites, a small quadrant as an inclinometer, and several protractors.

All these relics, except some small magnetic theodolites and protractors, are preserved by the Inō family. One of the small magnetic theodolites is in the possession of the Science Department of the Tōkyō Imperial University.

2 Maps

The various maps prepared by Inō at various times based on his survey-materials, have been discussed in the foregoing chapters. They are enumerated here again for review, together with a brief account of their contents and the places where they have been preserved.

(a) The maps of the south-eastern coast of Yezo and the Ōshū Highway drawn on the basis of the survey of 1800.

The large-scale map drawn on the scale of 1 : 43636 consists of 11 sheets for the Ōshū Highway and 10 sheets for the coast of Yezo—21 sheets in all. His skill in drawing displayed in this map does not compare with that shown in his later maps. Only 10 sheets for Yezo Island are preserved in the Tōkyō Imperial Museum. Nothing is known of the rest.

The small-scale map is drawn on the scale of 1:436363; the whole area surveyed is drawn on one sheet. He drew the meridians and parallels according to Sanson and Flamsteed's system. taking the meridian passing through the Astronomical Bureau in Asakusa, Yedo, as the prime meridian, and meridians and parallels are drawn at an interval of one degree. The length of a meridian degree is *27 ri*. The map is preserved at the Inō home and another original copy at the Tōkyō Imperial Museum.

(b) The maps of the eastern coast of the Main Island and the Ōshū Highway drawn on the basis of materials obtained by the survey of 1801.

The large-scale map consisting of 32 sheets drawn on the scale of 1:36000. This has been lost sight of.

The middle-scale map drawn on the scale of 1:216000 is based on the materials of the survey of 1800 in addition to those of the survey of 1801. This is made up of two sheets, on which the south-eastern coast of Yezo is also drawn. The length of one meridian degree is 28·2 *ri*; meridians and parallels are drawn by Sanson and Flamsteed's system at an interval of one degree. The meridian passing through Kuroye-chō, Fukagawa, Yedo, is drawn to intersect the parallels of latitudes perpendicularly and is taken as the prime meridian. The two sheets are preserved at the Inō home, but a half sheet is damaged.

The small-scale map drawn on the scale of 1:432000 on one sheet for the area and also the south-eastern coast of Yezo Island. This was simply a reduction of the middle-scale map to one half, with some omission. The whereabouts of this map is not known, but the manuscript of explanatory notes on it is still preserved at the Inō home.

(c) The maps of the coast-line between Aomori in Mutsu and Imamachi in Echigo, and the highways in Dewa and Echigo, based on the survey of 1802.

They must have been of three sizes—the large-scale, the middle-scale, and the small-scale—but documents on the subject are lacking, nor have the maps themselves been located.

(d) The maps of the coast-line of the Main Island east of Owari and Echizen drawn on the materials of the surveys dating from 1800 to 1803.

The large-scale map drawn as usual on the reduced scale of 1:36000 consists of 69 sheets for that part of the Main Island; no map for Yezo Island was made. The map is preserved entire at the Inō home.

The middle-scale map drawn on the scale of 1:216000 consists of 3 sheets including Yezo Island. Meridians and parallels are marked by Sanson and Flamsteed's system for each degree, and the meridian passing through Kuroye-chō in Yedo is taken as the prime meridian and is made to intersect the parallels at right angles. On the margin of the map is given a list of the latitudes and the successive distances of 300 important points along the route. Because the form of this map was most convenient for daily use, and the size was not too great, it was eagerly demanded by the dignitaries of these days. A number of copies were made. Two copies are preserved at the Inō home.

The small-scale map drawn on the scale of 1:432000 consists of one sheet for the whole area surveyed. The form does not differ from that of the middle-scale map, though it is somewhat abbreviated. The list of the latitudes and distances is not given. Explanatory notes by Inō, a preface by Kageyasu Takahashi, etc. are given on it. Because of its greater handiness than the middle-scale map, it seems that a number of duplicates were prepared, though only one is extant. Besides, not a few copies were made at a later date. The Geographical Society, Count Sano and others are in possession of one each. None of these, however, is dexterously copied.

(e) The maps of the coast of the Chūgoku region, and of the Tōkai Highway, based on the survey of 1805–1806.

The large-scale map was drawn on the scale of 1:36000. Neither the map nor accurate documents regarding it are extant. It is not even known how many sheets composed it.

The middle-scale map, drawn on the scale of 1:216000, consists of two sheets. The meridians and parallels are marked as usual, the prime meridian being the one passing through the site of the Calendaric Office in Kyōto and intersecting the parallels perpendicularly. Two duplicates are preserved at the Inō home; one has no meridians or parallels marked on it. Three other sheets which are alleged to be copies of it are preserved elsewhere.

The small-scale map was drawn on the scale of 1:432000, being just one half of the middle-scale map. Its location is not known.

The maps of some special localities. The region Inō surveyed in 1805–1806 comprised some places noted for fine scenery, such as Ama-no-Hashidate, Miyajima, and Lake Biwa. It seems that he made special maps of such localities and their vicinity on a properly reduced scale, and that he presented these to the Government, also distributing them among the high dignitaries and his acquaintances. A number of such maps are extant. Lake Biwa with its vicinity is drawn on the scale of 1:108000, Ama-no-Hashidate with its vicinity on the scale of 1:12000, and Miyajima with its vicinity on the scale of 1:36000; the description of the topographical features and other notations also vary in minuteness in accordance with the difference in scale. And at this time Inō also drew maps of some other special localities—Lake Hamana with its vicinity and Matsushima and its neighbourhood—based on the survey-work of the previous years. Of all the maps of special localities, Lake Biwa and its surroundings is the one that Inō most often reproduced. Three copies of it are preserved:

two at the Inō home and one elsewhere. One copy of Ama-no-Hashidate is preserved at the Inō home. Three copies of Miyajima are extant: one at the Inō home and two elsewhere.

(f) The maps of the coast of Shikoku and Awaji, the highways in Yamato, etc. drawn after the survey of 1808.

The large-scale map. The scale and the method of drawing were similar to those adopted previously for the large-scale maps. This is made up of 22 sheets in all; 17 sheets representing Awaji and Shikoku; 4 sheets the highway running from Ōsaka to Ise, and one sheet the Kiga Highway. The sheets for the coast of Awaji and Shikoku are lost, though the rest are preserved at the Inō home and a few fragments are also preserved elsewhere.

The middle-scale map of Awaji and Shikoku is drawn on the scale of 1:216000, on one sheet. The whereabouts of the original in which the meridians and parallels are marked is not known. A copy without them is preserved at the Kyōto Imperial University.

The small-scale map was drawn on the scale of 1:432000, or a half of the middle-scale map. Its whereabouts is not known.

(g) The provisional map of all Japan which Inō drew on the basis of his survey-materials of from 1800 to 1808 and also by referring to other materials then accessible.

This map hastily undertaken at the urgent request of the Shogunate Government, was completed and presented in August, 1809. Drawn on the scale of 1:864000, it represented on one sheet the whole country, including Yakushima and Tanegashima on the south-west and Hakodate and its vicinity on the north-east. The meridians and parallels were marked according to Sanson and Flamsteed's system, the prime meridian being the one passing through Kyōto, intersecting the parallels perpendicularly. In making this map, Inō had to depend on Sekisui Nagakubo's map for the Kyūshū region which Inō had not yet surveyed. The coast-line he had surveyed was given in red, and the rest in black. The whole length of the boundaries between provinces and the courses of rivers were indicated after referring to various provincial maps, and the more prominent towns throughout the country were also designated. On its margin an introduction by Kageyasu Takahashi, explaining how the map had been made, was given. A copy of the map was formerly preserved at the Inō home; after passing from one hand to another, it has been lost sight of.

(h) The map of the south-eastern coast of Kyūshū and some highways, drawn on the basis of the survey of 1809–1811.

The large-scale map was made almost on the same plan as the previous large-scale maps. It is evident that, besides the island of Kyūshū, the Nakasen Highway, the Sanyō Highway, etc. which Inō surveyed on his way there and back, were represented on the map. Only 21 duplicate sheets for Kyūshū are preserved at the Tōkyō Imperial Museum: it is not known whether the rest exist or not.

The middle-scale map represents the south-eastern region of Kyūshū on one sheet, and is drawn on the scale of 1:216000. A duplicate is preserved at the Tōkyō Imperial Museum. The parallels are marked, but not the meridians on this map. A copy of this map is preserved at the Kyōto Imperial University.

The small-scale map is drawn on the scale of one half the middle-scale map. This map is preserved at the Tōkyō Imperial Museum and also at the Chiba Middle School in Chiba Préfecture.

(i) The maps of the coast and highways of the remainder of Kyūshū, and some other highways drawn on the basis of the survey of from 1812 to 1814.

The large-scale map, drawn by the usual method, must have comprised the remainder of Kyūshū and the highways surveyed on his way out and back. Only a few sheets have been located, of which 7 sheets for Tsushima, Iki, Gotō, Hirado, Tanegashima, and

Yakushima are preserved at the Kyōto Imperial University, and 2 sheets, for Takayama in Hida Province and its vicinity and Ōmiya in Chichibu and its vicinity are in the possession of a relative of the Inō family.

The middle-scale map was no doubt one based on the survey-materials of 1809–1811, revised and enlarged to represent the whole of Kyūshū by the survey-materials of 1812–1814. This has been lost sight of.

The small-scale map was no doubt drawn on the scale of 1:432000, or a half of the middle-scale map. Neither has this been discovered.

(j) The maps of the eastern coast and the Seven Islands of Idzu, and various highways drawn on the basis of the survey of 1815–1816.

The special large-scale map for the Seven Islands of Idzu only, drawn on the scale of 1:12000, consists of 7 sheets. The general plan of describing various objects is similar to that usually adopted in the large-scale maps; only the indentations of the coast-line and the special features of mountains were much more accurately given. The map is preserved at the Inō home.

The large-scale map was no doubt drawn on the scale of 1:36000 for the whole area surveyed during the two years. Only one sheet for the Idzu Peninsula is preserved at the Inō home.

The middle-scale map drawn on the scale of 1:216000 for the peninsula and the Seven Islands on one sheet. The author has a duplicate of this map, on which the azimuth lines towards a great many mountains and islands are marked, though no meridian or parallel is given.

The small-scale map reduced to one half of the middle-scale is not located.

(k) The map of the streets of Yedo based on the survey of 1816.

It is true that Inō and his men had surveyed the principal highways in Yedo in the spring of 1815, connecting the Nihon-bashi, the point from which distances are measured, with the temporary termini at the main gates of the city, from which they had started their survey-work in the previous journeys, and then had made a map. This could not be called a map of the streets. The survey of 1816 furnished them for the first time with enough materials for preparing a map of the streets. The map was completed in the following year, and presented to the Shogunate Government. Yedo was drawn on the scale of 1 : 6000 on two sheets. On account of lack of materials, it was impossible to give the streets and residences in full detail, but the positions of various points were quite accurately given. It was really worthy of the name of a survey-map. The original map which had been preserved at the Inō home was donated to the Government some years ago and was in the custody of the Tōkyō Imperial University. It is a matter of great regret that it was reduced to ashes in the great earthquake and fire a few years ago. The Tōkyō Municipal Office has a copy, which had been made from the original some time before the calamity, and recently the reproduction of this copy was published.

(1) The complete map of Japan constructed by combining all the survey-materials of from 1800 to 1816 and those offered by Rinzō Mamiya.

The large-scale map drawn on the same plan as other large-scale maps is on the scale of 1 : 36000. It consists of 214 sheets, including Yezo Island. The original that had been preserved by the Shogunate Government, was handed on to the Imperial Government but was burnt up in the fire of 1873, and the one that had been kept at the Inō home was donated complete to the Government and was in the custody of the Tōkyō Imperial University, but this also was destroyed in the earthquake of 1923.

A copy of the latter made at a later date, however, is preserved in the Military Land-Survey Department.

The middle-scale map drawn on the same plan as other middle-scale maps is on the scale of 1:216000 and consists of 8 sheets. The meridian passing through the site of the Calendaric Office in Kyōto is taken as the prime meridian intersecting the parallels perpendicularly; the meridians and parallels are marked at an interval of one degree according to Sanson and Flamsteed's system. Azimuth lines are drawn for the more prominent mountains and islands. We have already called attention to the presence of the great drawing error which led to a large difference between the true longitudes of certain points and the longitudes measured referring to the meridians indicated on this map, and also to the presence of a considerable difference between the true azimuths and the azimuths measured on this map. The originals that had been in the custody of the Shogunate Government and the Inō family shared the fate of the large-scale maps. Fortunately, another has been found in the possession of Viscount Okōchi. Copies made at a later date are preserved at the Imperial Academy and elsewhere.

The small-scale map is one reduced to a half of the middle-scale with some omission. The marking of the meridians and parallels is similar to that adopted for the middle-scale map, and it is as erroneous in its projection as the other. It consists of 3 sheets. The originals that had been kept by the Shogunate Government and the Inō family were destroyed in fires and earthquakes. One which was formerly in the possession of an astronomer of the Shogunate is preserved elsewhere. As this map proved of much practical use in navigation before and after the Meiji Restoration, a great number of far-sighted men of different clans copied it. Consequently, not a few such copies are extant. Besides, this map was engraved with some modification and printed by the Kaiseisho about 1867.

The "Enkai-jissoku-roku" (The Record of the Coast-Survey), or *the "Yochi-jissoku-roku"* (The Record of the Survey of the Whole Country) which was presented to the Shogunate Government with the complete map of Japan consists of 13 volumes giving a full account of the latitudes of more than 1200 points on the coast-lines or highways Inō surveyed, the lengths of the coast-lines, and the lengths of the highways. It is accompanied by another volume giving prefaces and explanatory notes. The original of this book is lost, but copies are preserved by the Cabinet Library, the Imperial Library, and others. Though this was also printed in 1870 by the Daigaku Nankō, it is full of errors and some parts are missing.

The special small-scale map is one reduced to a half of the small-scale—namely, 1:864000. The whole country is given on one sheet. The meridians and parallels are drawn by the same method as the small-scale map. The only difference from the other is that it includes the whole lengths of all the boundaries between provinces and those of rivers, which Inō failed to survey and borrowed from other sources, and it also contains the more prominent mountains, towns and villages, and lakes and ponds which were not measured or even sighted from a distance by him. This map may be termed a revision of the map (g) provisionally drawn in August, 1809. The original of this map has not yet been located. A copy made at a later time is extant, but is by no means skilfully done.

The special small-scale map of a strange type which was abbreviated and modified from the foregoing is preserved at the Imperial Library. It consists of 3 sheets; the scale of reduction and the form of marking the meridians and parallels being exactly similar to those adopted for the special small-scale map. The positions and plan of drawing the surveyed lines, mountains, rivers, and the provincial boundaries on the area between Yaku-shima and Tanegashima on the south-west and the Tsugaru Straits on the north-east are also exactly similar to those adopted for the special small-scale map. The map of Yezo Island, however, is considerably revised from the various maps drawn in 1821; the number of surveyed lines among the mountains in the interior being much increased and the longitudes of various points not being similar to those given in the other maps. Besides, the eastern half of Kunajiri, Etrofu, and Urupp, which are not treated in the other maps, are given on this map, together with the whole of Saghalien and the area including the mouth of the Amur River. The mapping of the latter was evidently based on the map drawn by Rinzō Mamiya who personally traversed these regions, except the north-eastern coast of Saghalien. As Mamiya did not carry out proper survey-work, his positions are considerably erroneous,—namely, the whole island of Saghalien is placed between a little below 46° and a little below 51°, so that the point corresponding to 50° or the present boundary between Japan and Russia is given on the map approximately at 48°. The most conspicuous characteristic of this map is that it has no symbols for castle-towns, camps, or harbours; and the names of places on it, except those of provinces and counties, are somewhat fewer than those given in the special small-scale map, also they are given in *katakana* or the square style of the Japanese syllabary. We presume that this map was drawn by the men formerly under Inō, at the request of Kageyasu Takahashi, who desired to present it to Siebold, but failed to do so. That the names of places are given in *katakana* for the greater benefit of foreigners tends to prove that it is a relic of the great scandal.

(m) The maps enumerated above are those completed for some district or the whole of Japan. *The maps of small sections, "combined" maps, district maps, topographical sketches along the surveyed routes,* and various *map-drafts* which Inō made as preparations for drawing the above-mentioned maps, are also abundant. Though most of them have been scattered, those sheets of various sizes preserved at the Inō home reach several hundred. They are valuable as the originals of the maps, and also as materials for investigating the real state of the adjustment and composition of the maps. The Tōkyō Imperial University preserves several dozen small-section maps.

(n) Moreover, the copper-plate *sketch-map of the Japanese frontiers* published by Kageyasu Takahashi in 1809, covers not only the Japanese territories at that time, but also the peninsula of Korea, the north-eastern coast of China, the eastern coast of Siberia, and Saghalien and Kamtchatka. Takahashi compiled this map, referring to several maps and books known in Japan, China, and Korea, as well as to maps made in Russia, Portugal, etc. As the regions in this country which Inō had surveyed by that time were reproduced only on a reduced scale, but exactly as they were drawn by Inō, this map of Takahashi's, so far as these regions are concerned, may be called but another of Inō's maps.

3 Books and Documents

By the time his scholarship was attained, Inō was exclusively engaged in the task of surveying Japan and found no time for doing anything else. This accounts for the absence of his scientific papers on the results of the investigations. His chief works are mostly collections of the materials or results of his survey-work, documents concerning his survey-work, and tables of numerical values computed according to some definite formulæ. The principal books and documents may be mentioned here.

(a) *The "Sokuryō-yachō"* (The Field-Notes) in several hundred volumes. These have been almost completely destroyed.

Only a few pages, alleged to have belonged to them, are now preserved at the Inō home.

(b) *The "Santō-hōiki"* (The Record of the Azimuths of Mountains and Islands), of which 67 volumes are extant. Though some parts are thought to be missing, the volumes are preserved almost complete at the Inō home.

(c) *The "Hokkyoku-kōdo-sokuryōki"* (The Record of the Latitude Measurement) in several dozen volumes. Only one volume for the observation during 1802, is preserved at the Inō home.

(d) *The "Kōsei-shikōdohyō"* (The Table of Apparent Altitudes of Fixed Stars) in several volumes. One volume giving the results of observations made during 1802 and 1803, and another containing altitudes corresponding to the year 1813, are preserved at the Inō home.

(e) *The "Ichiri-rokubudzu-tōzai-no-keido-narabino-hokkyoku-kakoku-yori-no-chokuyenkeisa"* (The E.W. Co-ordinates and N.S. Co-ordinates on the Map Drawn on the Scale of 1:216000) in one volume. This is preserved at the Inō home.

(f) *The "Dai-Nippon-enkai-jissoku-roku"* (The Record of the Coast-Survey of All Japan), or *the "Enkai-jissoku-roku,"* in 14 volumes. Duplicates and printed copies are extant.

The contents of these books have already been discussed in the text, and need not be repeated here.

(g) *The "Sokuryō-nikki"* (The Survey-Journals) in 28 volumes. Beginning with the reproduction of the letter Yoshitoki Takahashi wrote to Inō concerning the survey of Yezo on March 10th, 1800, and concluding with the account of the return to Yedo on May 8th, 1816, of the survey-party sent to the Idzu region, the journals give all the record of Inō's survey-work during the 17 years. They are preserved complete at the Inō home. Another copy is in the possession of the Imperial Academy.

These survey-journals in 28 volumes were rewritten by Inō according to the original notes in which he and his men had made daily notations. The originals are also mostly preserved at the Inō home. Our success in knowing the real conditions under which Inō developed his survey-work is chiefly due to the existence in entirety of these rewritten survey-journals.

(h) *The "Yedo-nikki"* (The Yedo Diaries) in some volumes. From the fact that the journals were kept for the whole period that Inō was engaged in survey-work, it is easy to imagine that he was in the habit of keeping a diary from the time of his residence at Sawara till his last days. As for his diaries previous to the commencement of his survey-work, however, only one or two travels are left. Most of his diaries at Yedo, written after undertaking his survey-work, have been scattered; the only diaries still preserved at the Inō home are, one from December, 1806 to February, 1808, one from February, 1809 to September of the same year, and one from June, 1814 to January, 1815,—3 volumes all told.

(i) *The "Chikyū-sokuyenjutsu-mondō"* (Questions and Answers on the Method of the Distance Measurement on the Earth) in one volume. It is a pamphlet, or rather a collection of correspondence, exchanged between Inō and Shigetomi Hazama, on the method of computing the distance from either of the two survey-stations on the surface of the earth, considered as a sphere, to the bottom of a mountain on a great circle connecting the two points, and also of computing the height of the mountain by measuring the altitudes of the mountain at the two points and the distance between the two points. This amounts, after all, to the solution of a simple problem in trigonometry. This pamphlet is preserved at the Inō home.

(j) *The "Kokugun-chūya-jikoku"* (the Lengths of Day and Night in Different Places) in two volumes. This is a table of the hours of sunrise and sunset and of the lengths of day and night at various points of the country (at every 20′ between 34° and 36°, at each degree between 30° and 34°, and between 36° and 46°, and also at 20°, 25°, 50°, 55°, and 60°) computed for every 5 days throughout the various seasons. The lengths of day and night are computed in two ways: one for those limited by the rising and setting of the sun, and the other for those to which the so-called twilight duration is considered. This is preserved at the Inō home.

(k) *The "Taisūhyō-kigenjutsu-narabini-yōhō"* (The Principles and Uses of the Logarithmic Table) in two volumes. Of the two, the second volume is preserved at the Inō home. Though the first has been lost, and some parts of the contents are not quite clear, the cardinal points may be surmised. In this work, Inō expounds the method of computing logarithms devised by Chokuyen Ajima, the able scholar of Japanese mathematics, which may be summarized as follows: 10, the number whose logarithm is 1, is extracted to the tenth root to obtain the number whose logarithm is 0·1, and this is raised to the second, third, ...ninth power to obtain the numbers whose logarithms are 0·2, 0·3,....0·9, respectively. Then the number whose logarithm is 0·1 is extracted to the tenth root, to obtain the number whose logarithm is 0·01, which is raised to the second, third,....ninth power to obtain the numbers whose logarithms are 0·02, 0·03,.... 0·09, respectively. Such processes are repeated in order to construct a table of numbers whose logarithms are 1, 0·9, 0·8,....0·1, 0·09, 0·08,....0·01, 0·009, 0·008,....0·001, 0·0009, 0·0008,.... 0·0001, 0·00009, 0·00008,....0·00001, etc. Next, in order to obtain the logarithm for a given number A, A' the closest number to A, but a smaller, is selected from the table; by dividing A by A', B is obtained; then B' the closest number to B, but a smaller, is selected from the table; by dividing B by B', C is obtained; then

C' the closest number to C, but a smaller, is selected from the table; by dividing C by C', D is obtained. Such processes are continued until the number selected from the table sufficiently approaches 1, so that the logarithm for it is smaller than the smallest decimal unit of the logarithm demanded. Then the sum of logarithms for the number A', B', C', etc. is the required logarithm for the number A. In this book Inō explains the nature of logarithms and the principles of calculating them, and gives the logarithmic table for the numbers 1–100, 101–109, 1001–1009, 10001–10009, 100001–100009, 1000001–1000009, etc., which he computed according to Ajima's method, down to the tenth place of decimal. Inō perhaps planned to replace Ajima's original table by his own according to which the divisors in the processes in obtaining the logarithm for a given number are always integers, and division is rendered much simpler in order to avoid complication and error in computations. However, such a plan is neither explicity stated nor illustrated with examples in the second volume.

(1) *The "Kyū-katsuyen-hassenhyō"* (The Method of Computing Trigonometrical Tables). Though this work is not preserved at the Inō home, a piece of paper with this title, in Inō's hand, and given to Gendayū Uyeda of Amakusa about 1814, is still preserved at the Uyeda home. It explains that if a is the length of the circular arc of the radius r corresponding to a given angle, then the sine of the angle is obtained by

$$r\left\{\frac{a}{r}-\frac{1}{3\,!}\left(\frac{a}{r}\right)^3+\frac{1}{5\,!}\left(\frac{a}{r}\right)^5-\dots\right\}$$

and also that the cosine, the tangent, the cotangent, the secant, the cosecant, the versed sine, and the coversed sine are computed respectively by

$$\sqrt{r^2-\sin^2 a},\ \frac{r\sin a}{\cos a},\ \frac{r\cos a}{\sin a},\ \frac{r^2}{\cos a},\ \frac{r^2}{\sin a},\ r-\cos a,\text{ and } r-\sin a.$$

Thus the trigonometrical functions are expressed in the dimension of length.

(m) *The "Katsuyen-hassenhyō-kigenhō"* (The Principles of the Trigonometrical Tables). Although mentioned in his epitaph and other documents, this book has not been located. Whether what he wrote in the book was simply a recapitulation of the work of his predecessors or his own new design cannot be decided.

(n) *The "Bukkoku-rekishōhen-sekibō"* (In Confutation of 'The Indian Astronomy') in one volume. In this work Inō in the light of the latest theories based on observations at that time, confutes in Chinese the dogmatic unscientific opinions propounded in the "Bukkoku-rekishōhen" (The Indian Astronomy) written by the Priest Yentsū. The opinions in question were based on Buddhistic teachings, and aimed at the abuse and attack of western astronomical theories. Though absurd and nonsensical, the Bukkoku-rekishōhen was warmly received among some people, with some detriment to the advancement of new science. The date at which Inō wrote this is not clear, but it is apparent that he completed it not before the autumn of 1816, and that in spite of his wish to publish it he never succeeded in doing so. The manuscript is preserved at the Inō home.

(o) *The "Nippon-shokoku-hokkyoku-shutchido-oyobi-ritei"* (The Latitudes and Distances of Various Places in Japan) in several volumes. This is a summary of the latitudes and distances of the various points of the country surveyed by him. Only those for the region surveyed in 1800–1803 and those for Shikoku surveyed in 1808 are extant. The latitudes and distances given in these volumes are all collected after revision in the "Enkai-jissoku-roku."

(p) *The "Nippon-shokoku-hokkyoku-shutchido-oyobi-tōzai-keido,"* (The Latitudes and Longitudes of Various Places in Japan), in one volume. After the whole country was surveyed, this pamphlet was compiled for the latitudes of from two or three to ten towns in each province and for the longitudes of from one to three towns in each province. A table of latitudes and longitudes similar to this is also given in the special small-scale maps. Though the longitudes are given, down to the second, as though they had been accurately computed, they were mostly measured, on the map of the whole country with errors introduced in its drawing. They do not at all indicate the proper results of surveying.

(q) *The "Shikoku-oyobi-Kyūshū-no-ichibu-sokuryō-tōzai-oyobi-nambokudo"* (The E. W. and N. S. Distances in the Survey of Shikoku and Part of Kyūshū) in two volumes. This gives the E. W. and N. S. component distances between successive astronomical stations, obtained as a result of gradual mapping according to the material of land-surveying, the N. S. component distances immediately obtained by astronomical observation, and the E. W. and N. S. component distances determined by comparing and correcting the results of these two kinds—namely, the framework in constructing district maps. It seems evident that Inō made such record for all the regions he surveyed, but the two volumes here mentioned are all that are kept at the Inō home.

(r) *Miscellaneous Notes* in several volumes. Of all the numerous note-books in which he made some computations and wrote memoranda, only one has been preserved at the Inō home.

(s) *A Guide to the Chūgoku Route and the Kyūshū Highway* in one volume.

(t) *A Guide to the Kyūshū and the Tōkai Coasts* in one volume. These books give in order the names of the villages, which must be part of those abstracted from the original survey-book for convenience in putting geographical names on the maps.

(u) *The Castles and Camps of Various Provinces* in one

volume. This gives the localities of the castles and camps of the Daimios at the time of Inō's surveying.

(v) *Travels of Ōshū in the 7th Year of An-yei* in one volume.

(w) *Travels to the Kwansai Region in the 5th Year of Kwansei* in two volumes.

The former is a record of his trip to Matsushima in Ōshū in 1778, on which he was accompanied by his wife. The latter is of another trip made in 1793, accompanied by several relatives and acquaintances, to Ise to visit the Great Shrine, and to various noted places near Kyōto and Ōsaka. Neither is concerned with scientific discussions, but the former carefully gives the official distances between the successive posts while the latter gives the azimuths of mountains and islands roughly sighted from various points which commanded distant views, and the latitudes of some points, which serve to prove Inō's interest in geodesy even before he studied in Yedo. These are kept at the Inō home.

There also exist some documents which Inō himself did not write, but which were written by his friends or pupils, recording, the results of his observation, or expounding his method of surveying; and also some tables constructed for him at the Astronomical Bureau or elsewhere, in order to facilitate his surveying and computation.

(a′) *An Observation Record of the Solar and Lunar Eclipses by Tadataka Inō* in one volume. This is a record of the observation of some solar and lunar eclipses by Inō, which was compiled by Shigeyoshi (Jūshin), son of Shigetomi Hazama, with whom Inō himself or Yoshitoki Takahashi corresponded concerning the results. The solar eclipse of Nov. 8th, 1793, the lunar eclipse of Nov. 22nd of the same year, the solar eclipse of Apr. 24th, 1800, and the lunar eclipse of Aug. 22nd, 1812, are dealt with. It goes without saying that these are not all of the solar and lunar eclipses observed by Inō.

(b′) *The "Inō-Tōga-Sensei-ryū-ryōchi-denshūroku,"* (The Record of Training in Land-Surveying of Master Tōga Inō's School) in two volumes. This was compiled by Shin Watanabe (Keisuke Ogata), a pupil of Inō, who says in his introduction:—

". . . . Towards the end my master advised, 'There are hereditary professional scholars of astronomy, but none yet in the method of land-surveying. Should men be desirous of mastering my method after I die, show this book to them for their instruction'. . . . ."

These words clearly indicate the origin of the book which is the only document of Inō's method of land-surveying handed down to posterity. The book consists of only thirty odd sheets. The measuring-rope, the magnetic compass, the levelling-quadrant, the method of traverse, the protractor and the scale, the deformation of drawing-paper, the method of mapping, the method of intersection, and illustrations of the measuring-instruments are given under separate headings. However, no mention whatever is made of the method of determining longitude or latitude by astronomical observation. Elementary trigonometry and the trigonometrical tables are given as appendices. The former expounds the method of calculating the three other elements by the given three elements of a plane triangle, while the latter is the tables of the trigonometrical functions down to fifth decimals for each 5′. It seems that this work, planned for publication, was never really published. Though a number of written copies exist, not a printed one has been found.

(c′) *Tables of Eclipses and Occultations of Jupiter's Four Satellites* in several volumes. These are tables which the members of the Astronomical Bureau, at the request of Shigetomi Hazama and Kageyasu Takahashi, specially prepared and offered to Inō to assist him in his measurement of longitude, after computing the times of the eclipses and occultations of Jupiter's satellites. It is evident that these tables were constructed for many successive years perhaps after 1804, but only two volumes, for the period from the vernal equinox of 1805 to November, 1806, are preserved at the Inō home.

The tables give, in the local true solar time of Yedo, the times of the eclipses and occultations of Jupiter's four satellites to be observed at the Astronomical Bureau in Yedo. As we convert the times to Greenwich time and compare them with those given in the British Nautical Almanac of the same period, the times in the two tables for the first satellite generally coincide, with a difference of only 1 min., but as to those for the other satellites, the two tables sometimes differ as much as 10 min.

(d′) *The 10-Place Logarithmic Table for Numbers 1–100000*, complete in 5 volumes.

(e′) *The 7-Place Logarithmic Table for Numbers 1–10000*, complete in one volume.

(f′) *The 15-Place Sine and Cosine Tables for 0°–2° to Each Minute*, complete in one volume.

(g′) *The 8-Place Logarithmic Tables of Trigonometrical Functions, to Each Minute*, complete in one volume.

(h′) *The 7-Place Trigonometric Tables, to Every 1°/10*, complete in one volume.

(i′) *The 5-Place Trigonometric Tables, to Every 5′*, complete in one volume.

(j′) *The 5-Place Traverse Table*, complete in one volume. Two copies are extant.

These tables were of some practical use to Inō. Above all, the one most frequently used in reducing survey-materials was no doubt the traverse table. This was constructed for the purpose of simplifying the computations for obtaining the E. W. and N. S. component lengths by multiplying the length of each segment of the surveyed line by the sine and cosine of its azimuth. The

E. W. and N. S. component lengths for the length 9 *ken*–60 *ken* and for the azimuths 0°–90° are tabulated at an interval of 3 *ken* and of 5′. The tabulated numbers are given in 5 figures at the rate of 1 *bu* for 1 *ken*—namely, on the reduced scale of 1:600.

昭和七年五月十日印刷
昭和七年五月十五日發行

TADATAKA INŌ

定價金六圓

著者　大谷亮吉
姫路市五軒邸九六

譯者　杉村一枝
東京市外世田谷町下北澤二四五

發行者
發行所　岩波茂雄
東京市神田區一ツ橋岩波書店

印刷者
印刷所　島連太郎
東京市神田區美土代町三秀舍

大和會藏版

www.ingramcontent.com/pod-product-compliance
Lightning Source LLC
LaVergne TN
LVHW101612210926
848480LV00008B/132